Using Graphic Touch Screens and
SD Cards with Arduino

By

David Leithauser

Copyright © 2016
David Leithauser

All right reserved

Table of contents

Introduction..3
Hardware Aspects of Touch screens.................................5
Basic Software Information...13
Displaying Text and Numbers...15
Text Menus...23
Text With Scrolling..32
Text Menus With Scrolling..41
Numeric Keypads...46
Buttons..58
Histograms and Sliders..65
Plotting Data Like an Oscilloscope...................................85
Changing Screens...95
Using the Micro SD Card...111
Downloading the Files and Contact the Author..............132

Introduction

Arduinos are mini computers on a circuit board designed primarily for controlling electronic circuits. They have analog and digital inputs and outputs that allow them to input data and turn things on and off. For some applications (like building a robot), no direct interface with humans, such as a keyboard or display screen, is necessary. For some applications, however, you will want to display information and also input commands. For example, you may be designing a sensor device to monitor sensor readings (tricorder, anyone?). In this case, you will want to have a display. You may also want to input instructions or information into the Arduino. In that case, you will need some kind of buttons, menus, or analog control devices like volume controls.

You can have an LCD display to read information and some switches and knobs, but why not combine everything into one handy input/output device? Touch screens are perfect for this. They can display information as text or graphic. You can also cause them to display buttons or other graphic symbols that you can touch to make selections. One device can handle the entire human/Arduino interaction.

For this book, I will be working with the Seeed (that is not a misspelling, there are three e's.) Studio 2.4" 240 x 320 pixel touch screen Version 2 (http://www.seeedstudio.com). This is the best documented and most consistent touch screen on the market. However, if you are familiar with other brands of touch screens, you can still apply the basic code in this book. The libraries for each brand of touch screen might use slightly different commands for such things as initializing the screen, printing text to the screen, drawing rectangles or circles on the screen, and so on. However, if you simply substitute the appropriate command used by your screen for the one used in the sketches in this book, such as myGLCD.InitLCD() or Tft.init() for Tft.TFTinit(), the basic program concepts used in this book should work. I have written the code with most of the values, such as screen size, defined at the beginning of the sketch to allow for easy modification, and the purposes and functioning of most lines of code are explained step by step so you can understand what everything does and how to modify it to your needs.

I will start out by addressing some of the problems with connecting touch screens to Arduinos due to a bit of a design flaw in the shields. I will then go on to give sketches (Arduino talk for software programs, commonly referred to as apps these days) for creating text menus, buttons, sliders, and other controls that you are probably familiar with in normal windows programs. I will also discuss outputs like text (both static and scrolling), graphs and histograms for output. There is also a chapter on creating a system for swapping screens, so you can have more than one screen of controls or information within a sketch. Finally, I will show how to use the SD (or microSD) card that some touch screens have built in. The techniques used in that chapter should work on SD card readers that come separate from touch screens too. In the Appendix, I will provide information on how to download the sketches in this book to save you typing and how to contact me for questions.

Chapter 1

Hardware Aspects of Touch screens

I mentioned a design flaw in touch screen displays. The problems is that touch screens fit directly onto the Arduino and generally fill almost every pin on the Arduino Uno, even ones that the touch screen does not require an electronic connection to. Figure 1.1 shows a touch screen mounted on an Arduino Uno.

Figure 1.1

If you simply plug the touch screen into the Arduino the way it is designed, you will not be able to access any of the Arduino pins. This is fine if all you want to do is create a video game or something like that, but the main purpose of an Arduino is to input analog and digital data and control electronic and electrical devices. You want to be able to connect things to the Arduino using any pins that the touch screen does not require.

There are several ways to do this. One is to use a larger Arduino than the Uno, such as the Mega. Figure 1.2 shows the touch screen on a Mega.

Figure 1.2

As you can see, there are plenty of pins holes accessible. On the left, you can see analog 8 through 15 are easily accessible. Actually, analog pins 6 and 7 are available under an overhang. You can bend wires at a 90 degree angle and insert those under the touch screen. You can see digital pins 14 through 21 on the right, some of which serve extra functions like RX and TX. At the bottom of the picture you can see digital pins 22 through 53.

If you want to stick with the Arduino Uno, one solution is to mount a screw connector shield onto the Arduino and then put the touch screen on top of the screw shield. Screw shields have pins on the bottom that plug into the Arduino, pin holes on top that you can plug other shields into, and connectors on the sides that you can connect wires to. Figure 1.3 shows the top and bottom of a screw shield. Figure 1.4 shows a close-up of the sides where you can insert wires.

Figure 1.3

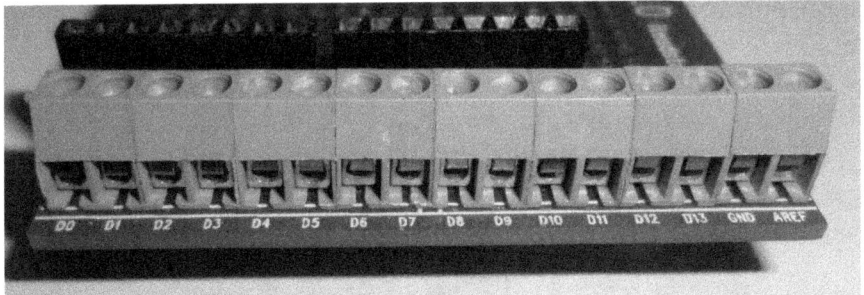

Figure 1.4

To access the holes on the sides, you loosen the screws on the top and the holes expand. You then insert the wires and rotate the screws back the other way to close the holes, clamping down on the wires. Thus, if you mount the screw shield on the Arduino and the touch screen on the screw shield, you have a way to make connections to the pins, and you can use this to connect to any unused Arduino pins.

There is one problem with this. The Seeed touch screen Version 2 does not just use the normal pin holes along the sides of the Arduino, such as the digital and analog pins. It also needs to connect to the two center pins of the 3X2 set of pins (referred to as ICSP or ISP) at the front of the Arduino, shown in Figure 1.5 with the pins that need to be connected circled.

Figure 1.5

If the touch screen is sitting directly on top of the Arduino, these pins fit into a socket on the bottom of the touch screen, shown in Figure 1.6.

Figure 1.6

8

When you separate the touch screen from the Arduino, you need to connect these separately. This means running wires from the two center holes of the socket to the corresponding pins. You can do this with male-female connector cables, such as those shown in Figure 1.7.

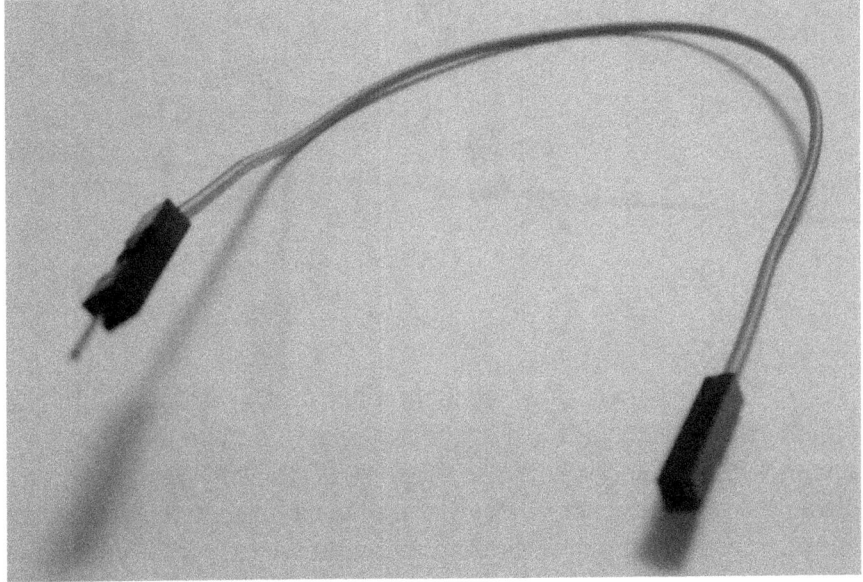

Figure 1.7

These are sold in ribbons (many wires side by side) on eBay and Amazon.com and are generally called Dupont male female connector cables for Arduino. You insert the male cable end into the socket and insert the pins from the Arduino into the female end of the cable. Unfortunately, there is not enough room between the touch screen and the screw shield or the screw shield and the Arduino for the ends of these cables, so you would need to bend the cable pin to about a 90 degree angle and bend the Arduino pins down. This is shown in Figure 1.8.

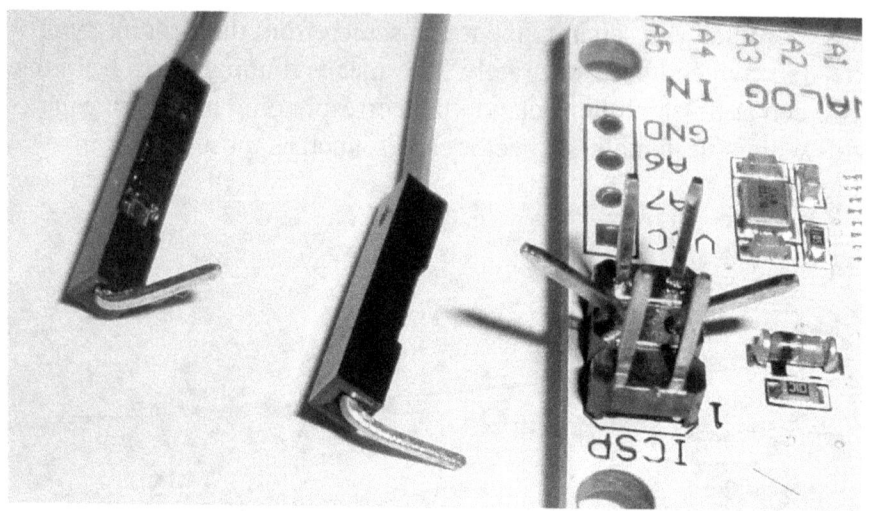

Figure 1.8

You can then plug the bent pins of the Arduino into the female end of the cables and the bent pins of the cable into the touch screen without the black plastic supports of the cable extending into the screw shield. Figure 1.8 shows the Seeed touch screen on top of the screw shield on top of the Arduino, with the extra pins connected by wires.

Figure 1.8

An alternative to using the male-female connector wire would be to use a 22 gauge insulated wire, insert one end of the wire into the socket hole and wrap the other end carefully around the Arduino pins, and possibly solder them. (You still probably need to bend the Arduino pins to make room to wrap the wires.) One more option would be to use female-female Dupont connector wires and insert one end of the 22 gauge wire into one end of the cable. That will allow you to insert the other end of the 22 gauge wire into the socket holes and put the other female end of the cable on the Arduino pins. The advantage to this is that the 22 gauge wire bends more easily than the pins at the male end of the connector wire, which can be brittle.

As long as you are using a screw shield, you can go one step further and use two shields. You can put one screw shield on the Arduino and put the touch screen on the second screw shield and connect each necessary connection of the two screw shields by wires to the screw terminals. You then also connect the two socket holes with the two Arduino pins as described above. This is shown in Figure 1.9.

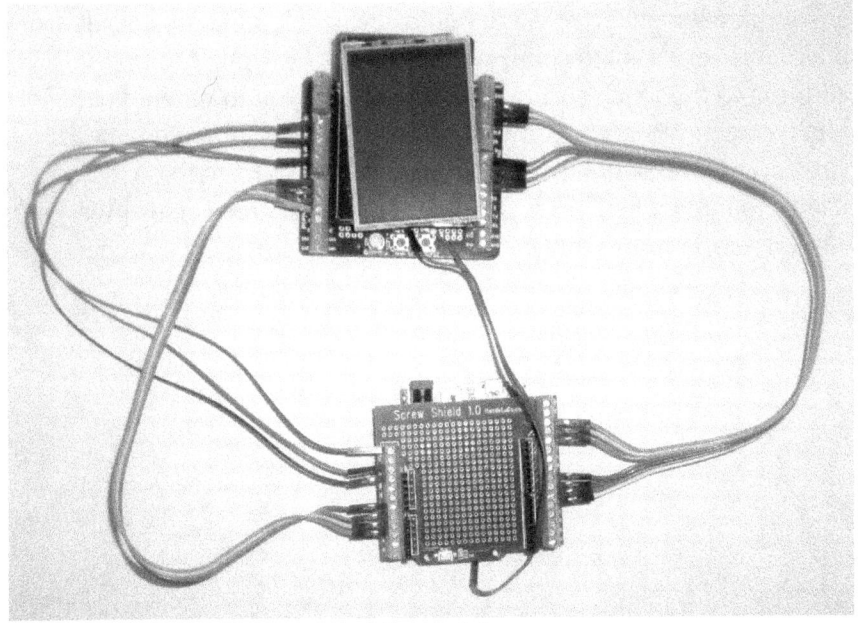

Figure 1.9

In order to do this, you must determine which pins need to be connected. For the Seeed touch screen, you need to connect D4 through D7, D11 through D13, A0 through A3, the GND and 5V connections, the reset pin (very important), and the two center pins of the 2x3 row of pins at the top of the Arduino. (Note: This connection list refers to version 2.0 and above of the touch screen. Version 1 requires D2 through D13 plus the analog pins A0 through A3, the reset pin, and ground and 5V, but does not seem to need the two extra separate connections described above.)

If you have any doubts about which pins need to be connected (if, for example, you are using a different type of touch screen), the best way to determine which terminals need to be connected would be to connect all the digital, analog, and power supply pins and then remove them one at a time, testing both the visual output and touch function of the screen after you remove each connection to make sure the screen is still fully functional. If an aspect of the screen stops functioning, reconnect that connection. Of course, you should not disconnect the power supply pin or the ground during this process. Sending power to the digital or analog inputs with either the 5V or ground disconnected could conceivably damage your touch screen or Arduino.

An advantage of using two screw shields like this, aside from making it very easy to connect other devices to your Arduino in addition to the touch screen, is that it allows you to mount your touch screen wherever you want on your device. If you do not use two screw shields in this way, you are forced to have the Arduino itself mounted wherever you want your touch screen. This gives you a good reason to use this method even if you are using a Mega.

Chapter 2

Basic Software Information

In order to use the touch screen with Arduino, you must download and install support files. Of course, you will first need the Arduino IDE, the program that will compile your code and install it in the Arduino itself. You can download that from
https://www.arduino.cc/en/Main/Software
Note: The above link, as of the writing of this book, downloads version 1.6.10 of the Arduino IDE. I found this version to be a bit buggy, especially when run on older operating systems. In particular, it had trouble compiling the code in this book running on an XP if you selected Arduino/Genuino Uno as your board. The sketches in this book were written using version 1.6.5. If you want the option of downloading older versions of the Arduino IDE, go to
https://www.arduino.cc/en/Main/OldSoftwareReleases
The library for the Seeed Studio touch screen can be downloaded from the Web page
https://github.com/Seeed-Studio/TFT_Touch_Shield_V2
and
https://github.com/Seeed-Studio/Touch_Screen_Driver
Note: If you have an older version 1 of the screen, you need
https://github.com/Seeed-Studio/TFT_Touch_Shield_V1
instead of the V2. However, this library seems to have some glitches that seem to prevent the touch feature from working properly. In particular, no matter where I touch the screen, I get an X and Y coordinate reading of -1. Thus, you can detect a screen touch but not where it was touched. I have been unable to determine the problem.

On these library download pages, click on the "Clone or download" button, and then on the "Download ZIP" option. Once you have the ZIP file downloaded, you can use the Arduino IDE to install it, provided you have at least version 1.6.5. Run the Arduino IDE. Click on the "Sketch" menu at the top, then go down to "Include library" option, then click on the "Add ZIP" option. On the screen that appears, move to the folder that you downloaded the ZIP file to. Find the ZIP file there, and double click on it. The file names should be TFT_Touch_Shield_V2-master.zip and Touch_Screen_Driver-master.zip. Of course, the file names can change after this book is published, so if you do not find that file, look for something similar.

In order to use the MicroSD card, you may also need to download the library for that. You can find that at
https://github.com/adafruit/SD
However, that is often included in the Arduino IDE library, so that may not be necessary.

If you do install this library separately, use the same procedure to download and install that file as the touch screen libraries above. The file name will probably be SD-master.zip.

In order to use these libraries, you must start your sketches with the proper include statements. For touch screen version 2, these are
#include <TFTv2.h>
#include <SPI.h>
#include <SeeedTouchScreen.h>
The SeeedTouchScreen.h file is actually only needed if you will be using the touch feature. That is, you do not need it if you are just displaying text, as we will be doing in the next chapter. However, that is certainly an unusual situation for a touch screen.

For version 1 of the touch screen, the includes are
#include <TouchScreen.h>
#include <TFT.h>

If you are going to be using the SD card, you also need to add #include <SD.h>. Some authors also use
#include <stdint.h>
However, this appears to be automatically included within the Arduino IDE code itself in newer versions, so it is usually unnecessary.

You also need to initialize the touch screen by putting a command in set setup routine. For the Seeed touch screen version 2, this command is Tft.TFTinit(). For version 1, it is Tft.init(). You can also turn on the backlight with the command TFT_BL_ON in the setup routine, although the backlight is turned on by default in many boards.

Chapter 3

Displaying Text and Numbers

The first and most obvious thing you need your touch screen to do is display text. This will be the main way that your screen will report information, and it is also necessary for text menus, buttons, and other input functions. Sketch 3.1 shows basic text output on the Seeed touch screen.

```
#include <TFTv2.h>
#include <SPI.h>

//Changeable settings
int screenHeight = 320;
int screenWidth = 240;
int textSize = 4;
int textSpacing = 5;
int textColor = CYAN;
int totalTextHeight = 7.5 * textSize + textSpacing;
int maxItems = screenHeight/(totalTextHeight);
// maxItems is maximum number of lines that will fit on screen

// Sample array of text to show
char* textItem[] = {"Item 1", "Item 2", "Item 3", "Item 4", "Item 5", "Item 6", "Item 7"};
int numItems=7;
int i;

void setup() {
  TFT_BL_ON; // turn on the background light
  Tft.TFTinit(); // initialize TFT library
  displayText();
}

void loop() {
  // No code in main loop for this example
```

}

```
void displayText() {
  Tft.fillRectangle(0, 0, screenWidth, screenHeight, BLACK);
  for (i=0;i<maxItems;i++) {
    if (i< numItems){
      Tft.drawString(textItem[i],0, i * totalTextHeight, textSize, textColor);
    }
  }
}
```

Sketch 3.1

Once you load the necessary libraries (the include statements), you have changeable variables or constants that will not be changed at the beginning of your sketch. These are all placed together at the beginning of the sketch rather than set in the individual routines so that you can easily find them and change them if they are not correct for your particular touch screen, or if you want to change something like the text size. It also insures that you do not accidentally have different values in different parts of the program. For example, if you had statements to draw text in several parts of the program using numbers instead of defined constants in each statement and you decided to change something like the text size, you might miss one and then the text would look different depending on which part of the program displayed it.

The screenHeight and screenWidth variables are simply the height and width of your touch screen in pixels. You would not change those unless you changed the touch screen. The textSize is the size of the text, and you can change that if you want. The textSpacing is the vertical space between lines, and again you can change that. The textColor of course is the color of your text. The totalTextHeight is the total number of pixels from the top of one line to the top of the next line. The program will use this to calculate where to put the text. The maxItems is the maximum number of lines your screen can display. You can see that totalTextHeight and maxItems are computed by the program, so if you change the text size or spacing you do not need to recompute these.

Next we input an array of seven text items to display. Of course, in an actual program, these might be variable, data the program is outputting to you from the Arduino inputs or something like a timer. However, it might also be a fairly constant set of items, like a text menu as will be discussed shortly. The variable numItems is simply the number of lines of text in your array. For some programs like this one, that number must be no more than maxItems. However, we will later get into scrollable text, in which case the number of lines of text in your array can be more than maxItems.

Note that although these were declared as variables, most of these will not change their value as this sketch executes. Therefore, you could declare them as constants by putting const in front of the declaration, such as changing

int screenHeight = 320;

to

const int screenHeight = 320;

Some newer versions especially suggest that you use constants for the char* when you assign a string literal to it. If you do not use constants, you can get error messages although the sketch will still compile. In this sketch, for example, you can convert

char* textItem[] = {"Item 1", "Item 2", "Item 3", "Item 4", "Item 5", "Item 6", "Item 7"};

to

const char* textItem[] = {"Item 1", "Item 2", "Item 3", "Item 4", "Item 5", "Item 6", "Item 7"};

However, you can only do this if the text does not change during execution of the sketch.

The code that displays the text is in the subroutine displayText(). Putting it in one subroutine allows you to change the contents of the text array in various parts of the program and then call the subroutine to display the text. The Tft.fillRectangle line simply clears the screen by filling it with a black box. The

for (i=0; i<maxItems; i++)

line will print the items in the text array, up to the maximum number of lines on the screen. The

if (i< numItems)

line makes sure that the program does not try to print more text lines than you have in the array. Thus the routine stops displaying text if either it reaches the end of the screen or it runs out of items in the array. The function

Tft.drawString(charVariable, horiznalPosition, verticalPosition, textSize, textColor);
displays to the screen the text in the char type variable at the position horiznalPosition, verticalPosition using text size given by the variable textSize in the color given by the color textColor. Therefore, the line
Tft.drawString(textItem[i], 0, i * totalTextHeight, textSize, textColor);
draws the text in the text array element i. The vertical location is i * totalTextHeight (i times the total height of each line of text). Note that i starts at 0, so the text starts at the top of the screen.

In this example, the subroutine to display the list of text is called from the setup routine. Since all this program does is display the text once, there is nothing in the loop section of the sketch. As we progress, there will be reasons to call the displayText routine from within the loop subroutine as the text changes.

The example in Sketch 3.1 displayed a simple text array. It can get slightly more complicated if you want to display text that is changed or manipulated within the program. This is because the drawString statement accepts a char or char* type variable, and these variables are hard to perform manipulations (like stringing several together) on. Such manipulations are better done with the String variable type. So, we have a conflict. Variable types char cannot be easily manipulated, and String type variables cannot easily be printed. The solution is to create a string to manipulate, and when you have finished getting the string you want, convert it to a char to display. This is a rather complicated multistep process, but Sketch 3.2 demonstrates it.

```
#include <TFTv2.h>
#include <SPI.h>

int screenHeight = 320;
int screenWidth = 240;
int textSize = 4;
int hPosition = 5;
int vPosition = 20;
int textColor = WHITE;
String textString;
int len;
```

```
int Measurement = 55;

void setup() {
 Serial.begin(9600);
 TFT_BL_ON;
 Tft.TFTinit();
 // Clear screen
 Tft.fillRectangle(0, 0, screenWidth, screenHeight, BLACK);
 // Assign string
 textString = "Data = " + String(Measurement);
 // Get length of string
 len=textString.length() + 1;
 //Define character set
 char DisplayableString[len];
 // Assign string to character set
 textString.toCharArray(DisplayableString, len);
 // Finally you can display the data
Tft.drawString(DisplayableString, hPosition, vPosition, textSize, textColor);
}

void loop() {
 // No code in main loop for this example
}
```
Sketch 3.2

First, to make the sketch general and easily changeable, we define some variables in the beginning. The variables hPosition and vPosition are the horizontal and vertical position to print the text at, and textColor is the color. The variable textString is the string you will manipulate. The variable len will be the length of the textString string plus 1. This is needed for the conversion process. Measurement is simply a variable that we will use in this example to represent some numeric value you obtained in the process of doing whatever you are doing, like taking a measurement on one of the Arduino analog inputs. For this example, we will just arbitrarily assign it a value of 55.

In the setup routine, we will assume that the value of Measurement has been determined, so textString is set to "Data =" plus this value with the statement

textString = "Data = " + String(Measurement);
This statement uses the String function to convert a numeric variable to a string variables and then concatenates that to the string "Data =" to form one string. This is one way of displaying numbers on the screen. The statement len=textString.length() + 1 finds the length of this string plus 1. The plus 1 is for a terminating 0 on the character string. The function string.length() is a built in function that determines the length of the variable string, so put your string name before the .length(). Next you need to define a char of the correct length. This is done with the
char DisplayableString[len];
statement, where DisplayableString can be any variable name. Note that len is used in the square brackets here to set the length of the char. Next the statement
textString.toCharArray(DisplayableString, len);
assigns the contents of the string variable textString into the char variable DisplayableString. Notice that you must once again give the length. Finally, you can now use the statement
Tft.drawString(DisplayableString, hPosition, vPosition, textSize, textColor);
to display the char variable DisplayableString at hPosition,vPosition with text size textSize in color textColor.

In Sketch 3.2, I showed one way to display numeric data on the screen, by converting it to a string with the String function and then using the drawString command. An alternative that can be simpler in many cases is to use the drawNumber command. This is very similar to the drawString command. The command
Tft.drawNumber(variable, x, y, textSize, textColor);
draws the value of the numeric variable variable at position x, y on the screen with size textSize in color textColor. Sketch 3.3 demonstrates this.

```
#include <TFTv2.h>
#include <SPI.h>

int screenHeight = 320;
int screenWidth = 240;
int textSize = 4;
int textSpacing = 5;
int textColor = CYAN;
```

```
int totalTextHeight = 7.5 * textSize + textSpacing;

int V1;
int V2;

void setup() {
 TFT_BL_ON;
 Tft.TFTinit();
 displayLabels();
}

void loop() {
 V1 = analogRead(A4);
 V2 = analogRead(A5);
 displayValues();
 delay(4000);
}

void displayLabels(){
 Tft.drawString("A4 =", 5, 5, textSize, textColor);
 Tft.drawString("A5 =",5,50,textSize,textColor);
}

void displayValues(){
 Tft.fillRectangle(120, 5, screenWidth - 100, totalTextHeight, BLACK);
 Tft.drawNumber(V1, 120, 5, textSize, textColor);
 Tft.fillRectangle(120, 50, screenWidth - 100, totalTextHeight, BLACK);
 Tft.drawNumber(V2, 120, 50, textSize, textColor);
}
```

Sketch 3.3

The subroutine displayLabels displays some text that goes before the numbers on the screen. This will only be needed to run once, provided you do not erase the screen, so the call for this subroutine can go at the end of the setup routine.

The subroutine displayValues displays the actual values of the variables V1 and V2, and must be called each time the variables change, or perhaps periodically. The trickiest part is positioning the variables directly after the labels on the screen. If you are putting the numbers at the end of the labels, the y position will be the same as the labels. I found by some trial and error that in this case, the x position of 120 puts the numbers a bit after the end of the label. Of course, the variables could actually go anywhere, such as below the labels, but in this example we are putting them immediately after the labels.

Note that we first use the fillRectangle command to draw a black rectangle where the numbers are to erase them. Otherwise, the numbers would keep writing on top of each other until you had a solid box of textColor. The x and y position of the rectangle should be the same as the text it is going to erase. The width should be the screen width minus the starting point so that it extends from the starting point to the end of the screen. The height should be the height of the text, which we calculated in the beginning of the sketch and called totalTextHeight. Thus, each time this subroutine is called, it erases the old numbers and draws the new values.

For this example program, we have the values of V1 and V2 being set within the main loop by reading the analog inputs A4 and A5. (Remember that analog ports 0 through 3 are used by the Seeed touch screen.)

Chapter 4

Text Menus

The previous chapter did not make any use of the touch feature of the touch screen. In this chapter, we will use the touch feature to turn the simple list of text in Chapter 3 into a touch menu. This is done in sketch 3.1.

```
#include <TFTv2.h>
#include <SPI.h>
#include <SeeedTouchScreen.h>

int screenHeight = 320;
int screenWidth = 240;
//Set minimum pressure to consider touch
int MinimumPressure = 50;
int textSize = 4;
int textSpacing = 5;
int textColor = CYAN;
int totalTextHeight = 7.5 * textSize + textSpacing;
// Maximum number of lines that will fit on screen
int maxItems = screenHeight/(totalTextHeight);
char* textItem[] = {"Item 1", "Item 2", "Item 3", "Item 4", "Item 5", "Item 6", "Item 7"};
int numItems  = 7;
int i;
int lineTouched;
 // Selection is which item was selected
int Selection = -1;
 // SelectionMade is was an item picked?
boolean SelectionMade = false;
 //init TouchScreen port pins
TouchScreen ts = TouchScreen(XP, YP, XM, YM);

 void setup() {
  Serial.begin(9600);
  TFT_BL_ON;
```

```
  Tft.TFTinit();
  displayText();
}

void loop() {
   // a point object holds x y and z coordinates.
   Point p = ts.getPoint();

   //map the ADC value read to into pixel co-ordinates
   p.x = map(p.x, TS_MINX, TS_MAXX, 0, 240);
   p.y = map(p.y, TS_MINY, TS_MAXY, 0, 320);
   // we have some minimum pressure we consider 'valid'
   // pressure of 0 means no pressing!
   // If pressure on screen
   if (p.z > MinimumPressure && p.z<10000) {
    lineTouched = p.y / totalTextHeight;
      if (lineTouched <= numItems-1) {
        Selection = lineTouched;
        SelectionMade = true;
     }
    }
     delay (250); // Prevent bounce. may vary time
   //Try to prevent continuous input
   if (SelectionMade && p.z==0) {
      Serial.println (textItem[Selection]);
     SelectionMade = false;
   }
}

void displayText() {
 Tft.fillRectangle(0, 0, screenWidth, screenHeight, BLACK);
 for (i=0;i<maxItems;i++) {
   if (i< numItems){
    Tft.drawString(textItem[i], 0, i * totalTextHeight, textSize,
textColor);
   }
  }
}
```

Sketch 4.1

Note that this is built on the code from Chapter 3. It contains the same displayText subroutine, and all the same constants, plus some additions. The MinimumPressure is the minimum pressure that you must exert on the screen for the sketch to consider it a touch. This can be set by you and does not change in the program. The lineTouched is which line of text on the screen was touched. The Selection variable is which item from the menu was picked. The value of lineTouched and Selection will be the same in this program, but in programs later in this book may be different when we get into scrollable text menus, so we will start out providing separate variables to allow for easier changes later. Note that a new include was added, SeeedTouchScreen.h, to provide the code for sensing the touch. This will, of course, be different for other touch screens. The code Serial.begin(9600); was added to the initialization to allow serial communication. This is not actually necessary for most programs, but I am using the serial output to allow the results of the menu selection to be sent to your computer for demonstration purposes. That is, this program will send the results of your selection to the computer to show you that it is working.

We now have something to put into the loop routine of the program, since this program will repeatedly check to see if the screen has been touched. The line
Point p = ts.getPoint();
tells the Arduino to read the touch screen to see if it is being touched and get the location and pressure of the touch. The lines
p.x = map(p.x, TS_MINX, TS_MAXX, 0, 240);
and
p.y = map(p.y, TS_MINY, TS_MAXY, 0, 320);
convert the raw touch information to screen coordinates, where p.z and p.y are the x and y coordinates of your touch.

The if statement that starts with
if (p.z > MinimumPressure && p.z<10000)
detects your touch. Notice that the pressure, p.z, must be greater than the threshold set by MinimumPressure. The requirement that p.z be less than 10000 is something I inserted because I found that sometimes the touch screen will suddenly start reading huge pressure, probably due to picking up electrical pulses in the air. The requirement that the pressure reading not be ridiculously high prevents that. The statement

lineTouched = p.y / totalTextHeight

converts the Y position of the touch into a line number of text. This value is then converted to a menu item number by the line

Selection = lineTouched

The flag SelectionMade is then set to true to show that a selection has been made. The statement delay (250) is to prevent switch bounce, the condition in which a switch can go on and off several times as you are pressing it as the contact is made.

    The if statment

if (SelectionMade && p.z==0)

then handles your selection. Notice that the program registered your touch in the previous if statement and set the SelectionMade flag to true. Once you release your touch, the pressure goes to 0 (p.z == 0) and the program acts on your touch. Note that the program does not act on your touch until you release the pressure. There are several reasons for this. One is that if the program acted when you were actually pressing on the screen, it would keep doing the action over and over as long as you were pressing on the screen. Requiring that you press and then release the screen prevents that. Second, it allows you to change your mind about which selection to made by sliding your finger (or stylus) to the correct line before you release the pressure. This will become clearer in the next program, which gives you visual feedback on what line you are touching.

    This particular example program simply sends the text of the line you pressed to the serial port with the line

Serial.println (textItem[Selection])

This is for demonstration purposes. In an actual program, you would have the Arduino act on your selection in some way. For example, suppose the first menu line was "Turn on" instead of "Item 1" and the second menu item is "Turn off" instead of "Item 2". Further, suppose you replace

Serial.println (textItem[Selection]);

with

if (textItem[Selection]=="Turn on") {digitalWrite(8, HIGH);
if (textItem[Selection]=="Turn off") {digitalWrite(8, LOW);

Pressing the first menu item on the screen would turn on whatever digital pin 8 was connected to and pressing the second menu line would turn it off. You would be controlling things with your touch screen. Of course, it can be other pins beside 8, but you must be sure that it is not one of the pins used by the touch screen, such as D4 through D7 and D10 through D13 for the Seeed touch screen.

Note that instead of comparing text strings, you could use code such as
if (Selection==0) {digitalWrite(8, HIGH);}
if (Selection==1) {digitalWrite(8, LOW);}
since the first line of text is line 0, the second line is line 1, and so on. This actually saves room in your program by using less code, but it is not as intuitive to program. As you get more experience, I suggest that you use this technique, especially if your code is long enough that memory space starts getting critical. However, when you start righting the program it might be easier to use the string comparisons to make sure you have the right lines.

We can improve on the touch screen by having a visual indication of which item you have touched. This will confirm that you have touched the right one and even allow you to correct it if you find that you are touching the wrong line by sliding your finger (or stylus) to the correct line before you release it. Sketch 4.2 shows how to do this.

```
#include <TFTv2.h>
#include <SPI.h>
#include <SeeedTouchScreen.h>

int screenHeight = 320;
int screenWidth = 240;
//Set minimum pressure to consider touch
int MinimumPressure = 50;
int textSize = 4;
int textSpacing = 5;
 //Color for normal text
int textColor = CYAN;
 // Color for selected text
int selectedTextColor = WHITE;
int printColor;
int totalTextHeight = 7.5 * textSize + textSpacing;
 // Maximum number of lines that will fit on screen
int maxItems = screenHeight/(totalTextHeight);
char* textItem[] = {"Item 1", "Item 2", "Item 3", "Item 4", "Item 5", "Item 6", "Item 7"};
int numItems=7;
```

```
int i;
int lineTouched;
int Selection=-1;
boolean SelectionMade = false;
//init TouchScreen port pins
TouchScreen ts = TouchScreen(XP, YP, XM, YM);

void setup() {
  Serial.begin(9600);

 TFT_BL_ON;
 // init TFT library
 Tft.TFTinit();
 Tft.fillRectangle(0, 0, screenWidth, screenHeight, BLACK);
  // Display the text
  displayText();
}

void loop() {
   // a point object holds x y and z coordinates.
   Point p = ts.getPoint();

   //map the ADC value read to into pixel co-ordinates

   p.x = map(p.x, TS_MINX, TS_MAXX, 0, 240);
   p.y = map(p.y, TS_MINY, TS_MAXY, 0, 320);
   // pressure of 0 (p.z = 0) means no pressing!
   // If pressure on screen
   if (p.z > MinimumPressure && p.z<10000) {
    lineTouched = p.y / totalTextHeight;
       if (lineTouched <= numItems-1) {
         Selection = lineTouched;
         displayText();
         SelectionMade = true;
       }
     }
     // Prevent bounce. may vary time
     delay (250);
    //Try to prevent continuous input
    if (SelectionMade && p.z==0) {
```

```
      // Take some action on selection
      Serial.println (textItem[Selection]);
      SelectionMade = false;
    }
}


void displayText() {
  for (i=0;i<maxItems;i++) {
    if (i< numItems){
      if (i == Selection){
        printColor = selectedTextColor;
      }
      else {
        printColor = textColor;
      }
      Tft.drawString(textItem[i], 0 , i * totalTextHeight, textSize, printColor);
    }
  }
}
```

Sketch 4.2

The first thing to notice is that the declarations
int selectedTextColor = WHITE;
and
int printColor;
have been added to the beginning of the program and this code segment has been added to the displayText subroutine.
```
      if (i == Selection){
        printColor = selectedTextColor;
      }
      else {
        printColor = textColor;
      }
```
and textColor has been changed to printColor in the Tft.drawString statement. These changes cause the program to draw the selected text line in a different color than the other lines. One other change you might notice is that the fillRectangle statement has been moved

from the displayText subroutine to the setup subroutine. The reason for this is that the displayText subroutine will be called frequently from within the program, not just from the setup subroutine, in order to change the color of the one line that has been selected. If the fillRectangle statement was still within the displayText subroutine, the entire screen would blink off briefly each time the text was rewritten. This is unnecessary, because when the displayText subroutine writes each line of text on the screen, it overwrites the text that was there before. Thus, when the program calls displayText in order to change the text color of some lines, you do not see a flicker.

Now, in addition to displayText being called from the setup subroutine, it is also called from within the code that detects that the screen is being pressed on, the part of the code that is within the
if (p.z > MinimumPressure && p.z<10000)
statement. This means that each time the program gets to this part of the program, the text will be rewritten as long as you are pressing on the screen. This is why we had to remove the screen blanking statement from the displayText subroutine, or you would see the screen clearing and being rewritten constantly as the program goes through the loop. Without this clearing occurring, event though the program is repeatedly rewriting the text, you do not see the text being rewritten.

So what good is this? As soon as you touch a line of text on the screen, it changes color from cyan to white. (Of course, you could pick other colors by setting textColor and selectedTextColor differently in the beginning of the program.) This tells you which line you have touched. If it is not the line you meant to touch, you can slide your finger down the screen without lifting it off the screen. The line of text that is set to a different color will change as you move your finger. When it gets to the right line, release the pressure on the screen. When you release the pressure on the screen, p.z will be 0 and the line
if (SelectionMade && p.z==0)
will take effect. Thus putting pressure on the screen sets SelectionMade to true, and removing it triggers the action. Since no action is taken until you have removed your finger, only the line you had your finger on when you released it will actually count. Note that once the action is taken SelectionMade is set to false again, so the action does not repeat.

With this code, the last line touched and released remains white on the screen. This shows you what you last did. However, if you want the color to go back to cyan after you release the pressure, just add

Selection=-1;
displayText();

after

SelectionMade = false;

Chapter 5

Text With Scrolling

Sometimes when you want to display text, there may be more text than you have room for on the screen. Of course, you can make the text smaller with the textSize variable, but that makes it harder to read, much harder to select the right line on a text menu, and you are still limited on how much text you can get on one screen even if you make it very small. The solution is to be able to scroll the text up the screen as you read it. Sketch 5.1 shows how to do this.

```
#include <TFTv2.h>
#include <SPI.h>
#include <SeeedTouchScreen.h>

int screenHeight = 320;
int screenWidth = 240;
int textSize = 4;
int textSpacing = 5;
int textColor = CYAN;
int totalTextHeight = 7.5 * textSize + textSpacing;
int maxItems = screenHeight/(totalTextHeight); // Maximum number of lines that will fit on screen
char* textItem[] = {"Line 1", "Item 2", "Item 3", "Item 4", "Item 5", "Item 6", "Item 7", "Item 8", "Line 9", "Item 10", "Item 11", "Item 12"};
int numItems=12;
int i;
int MinimumPressure = 50;
//New declarations start here
// Top item shown
int startItem = 0;
// Scroll up or down
int UpDown;
int rightMargin = screenWidth * .8;
```

```
//init TouchScreen port pins
TouchScreen ts = TouchScreen(XP, YP, XM, YM);

void setup() {
 Serial.begin(9600);
 TFT_BL_ON;
 Tft.TFTinit();
 Tft.fillRectangle(0, 0, screenWidth, screenHeight, BLACK);
 displayText();
}

void loop() {
    // a point object holds x y and z coordinates.
    Point p = ts.getPoint();
    //map the ADC value read to into pixel co-ordinates
    p.x = map(p.x, TS_MINX, TS_MAXX, 0, 240);
    p.y = map(p.y, TS_MINY, TS_MAXY, 0, 320);

    // we have some minimum pressure we consider 'valid'
    // pressure (p.z) of 0 means no pressing!
    if (p.z > MinimumPressure && p.z<10000) {
    //if touching upper right corner
       if (p.y < totalTextHeight && p.x > rightMargin){UpDown = -1;}
     // if touching lower right corner
       if (p.y > screenHeight - totalTextHeight && p.x > rightMargin) {UpDown=1;}
       // Prevent bounce. May vary time
       delay (500);
     }
    if (UpDown!=0) {
      if (p.z == 0) {
        startItem=startItem - UpDown;
        //Do not let top item be less than first item
        if (startItem<0) {
          startItem = 0;
          }
        //Do not let bottom item be more than last item
        if (startItem>numItems-maxItems) {
          startItem = numItems -maxItems;
```

```
      }
      UpDown=0;
      displayText();
    }
  }
}

void displayText() {
  Tft.fillRectangle(0, 0, screenWidth, screenHeight, BLACK);
  for (i=0;i<maxItems;i++) {
    if (i + startItem < numItems){
      Tft.drawString(textItem[i + startItem], 0, i * totalTextHeight, textSize, textColor);
    }
  }
}
```

Sketch 5.1

With this code, the program will print up to maxItems lines of text on the screen and then stop. You can scroll down the text but pressing and releasing the upper right corner of the screen. You can scroll down by pressing and releasing the lower right corner of the screen. Each press and release scrolls down or up one line.

This requires some changes to the displayText subroutine. The fillRectangle command, which clears the screen, has been moved back into the subroutine. This is because the drawstring command does not erase the previous text, so you could have residual text where the new text is printed. For example, it the line on the screen had been "Item 10" and you printed "Item 9" in that spot, the 0 would still be there. In fact, it would be even worse because the 1 would still be there, so the 9 would appear to have a line going up the middle of it.

If you want the region where touching the screen causes text to scroll to be visibly marked to make it easy to see where to touch, you can mark the regions with a small square by adding the following code immediately after the fillRectangle command that clears the screen.

```
Tft.fillRectangle(rightMargin, 0, screenWidth, totalTextHeight, BLUE);
```

Tft.fillRectangle(rightMargin, screenHeight - totalTextHeight, screenWidth, screenHeight, BLUE);

Of course you can control what color the squares are by changing the BLUE to some other color like RED. You obviously do not want to use BLACK, since that is the background color and it will not be visible. You also do not want to use the same color as the text. If the text is long enough, it will extend into the square and overwrite it. This is desirable, since the text is more important than the square.

A more important change is the addition of the startItem variable. This refers to what line is at the top of the screen. The statement

for (i=0; i<maxItems; i++)

counts through the lines on the screen, just like in previous sketches. However, the line

Tft.drawString(textItem[i + startItem], 0, i * totalTextHeight, textSize, textColor);

does not simply print item i on the screen, but i + startItem. This means that the first line of text displayed on the screen is not the first item in the text array, but startItem. The displayText subroutine then prints the following lines of text in the array, up to either the maximum number of lines on the screen or the last item in the array. Thus, by controlling the value of startItem, you can control which items in the text array are displayed.

Within the loop subroutine, the line

if (p.z > MinimumPressure && p.z<10000)

first checks to see if the screen is being pressed. The line

if (p.y < totalTextHeight && p.x > rightMargin)

checks to see if you are pressing the upper right corner. The p.y < totalTextHeight part checks to see if the touch is above the height of a line of text. The p.x > rightMargin part checks to see if touch is within the right margin of the screen. This value was set by the line

int rightMargin = screenWidth * .8;

You can vary the width of this area by changing the .8 value if you like. If these two conditions are met (the && means "and"), then UpDown is set to -1.

The line

if (p.y > screenHeight - totalTextHeight && p.x > rightMargin)

checks to see if you are pressing the lower right corner. If that is true,

the value of UpDown is set to 1. This is followed by a delay statement to prevent switch bounce. After that, the line
if (UpDown!=0)
checks to see if UpDown has been set to something other than 0 (!= means NOT equal to). If that is the case, the line
if (p.z == 0)
checks to see if the pressure has been released. If those two conditions are met, the line
startItem = startItem - UpDown;
adjusts the value of startItem up or down by subtracting UpDown from startItem. The program then checks to make sure that startItem is not too high or too low. The lines
if (startItem < 0) {
    startItem = 0;
}
reset startItem to 0 if it is less than 0. Otherwise, the displayText subroutine would be trying to display an array element, -1, that does not exist. The lines
if (startItem > numItems - maxItems) {
    startItem = numItems - maxItems;
}
are a little more complicated. They could just limit the maximum value of startItem to numItems by checking to see if startItem is greater than numItems, the last item in the text array. However, to prevent a lot of empty space on the screen, they limit the highest value of startItem to the last item in the array (numItems) minus the maximum number of lines on the screen. This means that the text will stop scrolling when the item on the bottom of the screen is the last line of text in the text array.

    Next the program resets UpDown to 0 since startItem has already been changed and we do not want it to keep changing each time the program hits this part of the code. Then the program calls the displayText subroutine to redraw the text on the screen with the new top line, startItem. The loop then repeats, waiting for you to touch the screen again.

    A more familiar way to scroll the text is to move your finger up or down along the screen. This is a bit trickier to implement on the touch screen because the Arduino touch screens require you to press on the screen, not just lightly touch them like the touch screens on most phones or tablets. This makes it a bit harder to move your finger smoothly along the screen. However, if you want to implement scrolling this way, Sketch 5.2 shows how to do this.

```
#include <TFTv2.h>
#include <SPI.h>
#include <SeeedTouchScreen.h>

int screenHeight = 320;
int screenWidth = 240;
int MinimumPressure = 50;
int textSize = 4;
int textSpacing = 5;
int textColor = CYAN;
int totalTextHeight = 7.5 * textSize + textSpacing;
int maxItems = screenHeight/(totalTextHeight);
char* textItem[] = {"Item 1", "Item 2", "Item 3", "Item 4", "Item 5", "Item 6", "Item 7", "Item 8", "Item 9", "Item 10", "Item 11", "Item 12", "Item 13", "Item 14", "Item 15"};
int numItems=15;
int i;
int startItem = 0;
int startTouch = -1;
float sensitivity = 4;

TouchScreen ts = TouchScreen(XP, YP, XM, YM);

 void setup() {
  TFT_BL_ON;
  Tft.TFTinit();
  displayText();
 }

void loop() {
   // a point object holds x y and z coordinates.
   Point p = ts.getPoint();
   //map the ADC value read to into pixel co-ordinates
   p.x = map(p.x, TS_MINX, TS_MAXX, 0, 240);
   p.y = map(p.y, TS_MINY, TS_MAXY, 0, 320);

   // we have some minimum pressure we consider 'valid'
```

```
  // pressure of 0 means no pressing!
  if (p.z == 0) {
     // Set startTouch something it will never actually be if screen touched
     startTouch = -1;
     }
  if (p.z > MinimumPressure && p.z<10000) {
     // Prevent bounce. may vary time
     delay (200);
     if (startTouch == -1) {
       startTouch = p.y;
       }
     if (p.y - startTouch > totalTextHeight/sensitivity ) {
       if (startItem >0) {
         startItem = startItem - 1;
         startTouch = -1;
         displayText();
        }
      }
     if (startTouch - p.y > totalTextHeight/sensitivity ) {
       if (startItem < numItems - maxItems) {
        startItem = startItem + 1;
        startTouch = -1 ;
        displayText();
         }
        }
       }
     }

void displayText() {
  Tft.fillRectangle(0, 0, screenWidth, screenHeight, BLACK);
  for (i=0;i<maxItems;i++) {
   if (i + startItem < numItems){
     Tft.drawString(textItem[i + startItem], 0, i * totalTextHeight, textSize, textColor);
    }
   }
 }
```

Sketch 5.2

We have a few new variables. The most important is startTouch, which tells the program where you first touched the screen. It is initially set to -1, which is a value that it cannot have if you have touched the screen since screen positions start at 0. Thus, if startTouch equals 0, it indicates that the screen is not being touched. Note that the first thing in the loop, once the program has read the pressure and x and y coordinates if the screen is being touched, is a statement

if (p.z == 0) {
  startTouch = -1;
}

Thus startTouch is always reset to -1 as soon as you remove your finger from the screen. This is followed by a set of events that occurs if (p.z > MinimumPressure && p.z<10000), in other words, if the screen is being touched. First (after a delay statement to avoid switch bounce), if startTouch is currently -1 it is set to the y coordinate of your touch by the statement

if (startTouch == -1) {
  startTouch = p.y;
}

That is, the program records the point at which you first touch the screen after you were not touching it.

Then, the program starts checking to see if you have moved your finger vertically since you started touching the screen. The statement

if (p.y - startTouch > totalTextHeight/sensitivity)

checks to see if you have moved you finger down significantly while touching the screen, since p.y increases as you touch the screen lower. If this distance is greater than totalTextHeight/sensitivity, this indicates that you have moved your finger a significant amount upward since you started touching the screen. Note the measure of what is significant is expressed in terms of the height of text. This is divided by a sensitivity setting. The greater the value of the sensitivity variable, the less distance you have to move your finger for it to be considered significant enough for the text to scroll. The value of sensitivity is set in the beginning of the program, so you an easily tweak it if you like.

If the program determines that you have moved your finger down at least the predetermined amount, it prepares to move the text down. First, it makes sure that the first item in the text array is not

already at the top of the screen with the line
if (startItem >0)

If this condition is met, startItem is decreased by 1, moving the text up the screen. After this, startTouch is reset to -1. This is necessary because if it was not, the value of startItem would continue to be decreased each time the program looped around to this portion, causing the text to scroll rapidly even though you were holding your finger still. The displayText subroutine is then called to display the text. That is the end of the code to see if you are moving your finger down.

After testing to see if you have moved your finger down, the program checks to see if you have moved it up. The statement
if (startTouch - p.y > totalTextHeight/sensitivity)
checks to see if the current position, p.y, is less than the start position, startTouch, by more than totalTextHeight/sensitivity. If so, it prepares to move the text up. First, it makes sure that the text is not already moved up to a point where the last text item is already at the bottom of the screen, in which case scrolling the text up farther would simply reveal blank lines. This is done by the line
if (startItem < numItems - maxItems)

If all these conditions are met, startItem (the line of text displayed at the top of the screen) is increased by 1. As in the previous section, startTouch is reset to -1 to prevent the text from scrolling rapidly even though you were holding your finger still. The displayText subroutine is then called to display the text.

Chapter 6

Text Menus With Scrolling

Now that we have text menus and scrollable text, let's combine them in case we want to have a text menu too long for the screen. Sketch 6.1 shows how to do this.

```
#include <TFTv2.h>
#include <SPI.h>
#include <SeeedTouchScreen.h>

int screenHeight = 320;
int screenWidth = 240;
int textSize = 4;
int textSpacing = 5;
int textColor = CYAN;
int totalTextHeight = 7.5 * textSize + textSpacing;
int maxItems = screenHeight/(totalTextHeight); // Maximum number of lines that will fit on screen
char* textItem[] = {"Item 1","Item 2","Item 3","Item 4","Item 5","Item 6","Item 7","Item 8","Item 9","Item 10","Item 11","Item 12","Item 13","Item 14","Item 15"};
int numItems=15;
int i;
int itemNumber;
int startItem = 0; // Top item shown
int previousStartItem = 0;
int UpDown; // Scroll up or down
int rightMargin = screenWidth*.85; //New declarations start here
int lineTouched; // Line on screen touched
int Selection=-1; //Item picked from list, initial to less than any item
boolean SelectionMade = false; //Was an item picked?
int selectedColor = WHITE;
int printColor;
int MinimumPressure = 50; //Minimum pressure to consider touch

TouchScreen ts = TouchScreen(XP, YP, XM, YM); //init
```

TouchScreen port pins

```
void setup() {
 Serial.begin(9600);
 TFT_BL_ON;
 Tft.TFTinit();
 displayText();
}

void loop() {
   // a point object holds x y and z coordinates.
   Point p = ts.getPoint();

   //map the ADC value read to into pixel co-ordinates
   p.x = map(p.x, TS_MINX, TS_MAXX, 0, 240);
   p.y = map(p.y, TS_MINY, TS_MAXY, 0, 320);

   // we have some minimum pressure we consider 'valid'
   // pressure of 0 means no pressing!
   if (p.z > MinimumPressure && p.z<10000) {
      if(p.y < totalTextHeight && p.x > rightMargin){UpDown = -1;}
      if (p.y > screenHeight - totalTextHeight && p.x > rightMargin) {UpDown=1;}
       if (p.x<rightMargin) {
        lineTouched = p.y / totalTextHeight;
         Selection = lineTouched + startItem;
         SelectionMade = true;
         displayText();
       }
      delay (500); // Prevent bounce. may vary time
    }
    //Try to prevent continuous input
    if (SelectionMade && p.z == 0) {
      Serial.println (textItem[Selection]);
      SelectionMade = false;
    }
    if (UpDown!=0) {
     if (p.z == 0) {
       startItem=startItem-UpDown;
```

```
      //Do not let start item be less than first item
      if (startItem<0) {
        startItem = 0;
        }
      if (startItem>numItems - maxItems) {
        startItem = numItems - maxItems;
        }
      UpDown=0;
      displayText();
     }
   }
}


void displayText() {
 if (startItem != previousStartItem) {
  Tft.fillRectangle(0, 0, screenWidth, screenHeight, BLACK);
  previousStartItem =  startItem;
  }
   Tft.fillRectangle(rightMargin, 0, screenWidth, totalTextHeight, BLUE);
   Tft.fillRectangle(rightMargin, screenHeight - totalTextHeight, screenWidth, screenHeight, BLUE);

   for (i=0;i<maxItems;i++) {
    if (i+startItem==Selection) {
     printColor = selectedColor;
    }
    else {
     printColor=textColor;
    }
     Tft.drawString(textItem[i + startItem], 0, i * totalTextHeight, textSize, printColor);
  }
}
```

Sketch 6.1

This sketch combines the text menu code from chapter 4 with the scrolling function in chapter 5. For this sketch, we have used the touch-the-right-corners method of scrolling the screen. Trying to slide the text using the swipe method without accidentally selecting an item can be tricky. To make it clearer where you can touch the screen to scroll instead of selecting an item, I have added a blue rectangle at the upper and lower right corners of the screen where you touch to scroll the text.

What we have done here is taken Sketch 5.1 and added in the lines that allow you to select an item. The main additional code is

```
if (p.x<rightMargin) {
   lineTouched = p.y / totalTextHeight;
   Selection = lineTouched + startItem;
   SelectionMade = true;
   displayText();
}
```

to select an item and

```
if (SelectionMade && p.z == 0) {
   Serial.println (textItem[Selection]);
   SelectionMade = false;
}
```

to act on the selection. Of course, as before, the action will probably be something more significant than simply sending the text to the serial port. You will probably want to replace
Serial.println (textItem[Selection]);
With something like

```
if (textItem[Selection]) == "Turn on") {
digitalWrite(8, HIGH);
}
```

Note that we have had to make some adjustments to Sketch 4.1. In that sketch, to select a line of text, you merely needed to set the Selection variable to the line on the screen, lineTouched. However, with scrollable text, the number of the element in text array does not necessarily match the line on the screen. Instead, we must set Selection to lineTouched + startItem, where startItem is the top line on the screen. This is why we have been preparing for this by using separate variables, lineTouched and Selection, even though until now they have been equal.

The displayText subroutine incorporates components of both Sketch 5.1 and Sketch 4.1, since it must both allow for startItem in deciding which lines of text to show and use Selection to determine the color of the text. Naturally, Sketch 6.1 must define all the variables for both of the other sketches at the beginning.

One additional feature you might have noticed is the addition of the variable previousStartItem. In the sketch to scroll text without being a menu, the only time the displayText subroutine was called was when the text was scrolling, and therefore it was necessary to clear the screen with the Tft.fillRectangle command to remove residual text. In this sketch, the displayText subroutine is called when either the text scrolls or the color changes because of a selection. We do want the screen to clear when the text scrolls, but having the screen clear when a selection is made is unnecessary because the printing of the same text in the same place, only a different color, exactly overwrites the previous text. Clearing the screen each time a selection is made introduces an annoying flicker to the screen. Replacing just

Tft.fillRectangle(0, 0, screenWidth, screenHeight, BLACK);
with
if (startItem != previousStartItem) {
  Tft.fillRectangle(0, 0, screenWidth, screenHeight, BLACK);
  previousStartItem = startItem;
}

causes the screen to clear only when the text scrolls. The if statement tests to see if the start item has changed. If so, it clears the screen and sets the previousStartItem to the current startItem.

Chapter 7

Numeric Keypads

An important input you will need for some Arduino projects is the ability to input numeric values. Sketch 7.1 shows how to do this with an onscreen numeric keypad.

```
#include <TFTv2.h>
#include <SPI.h>
#include <SeeedTouchScreen.h>

int screenHeight = 320;
int screenWidth = 240;
int MinimumPressure = 50; //Minimum pressure to consider touch

int textSize = 4;
int textHeight = 7.5 * textSize; // + textSpacing;
int textColor = BLACK;
int buttonWidth = 60;
int buttonHeight = 40;
int buttonHSpace = 10;
int buttonVSpace = 15;
int topSpace = 50;
char* buttonText[] =
{"1","2","3","4","5","6","7","8","9","E","0","C"};

int x;
int y;
int px;
int py;
int selection = -1;
int startTouch = -1;
String combined = "";
int lenCombined = 0;
long value;
```

```
TouchScreen ts = TouchScreen(XP, YP, XM, YM); //init
TouchScreen port pins

void setup()
{
  Serial.begin(9600);

  TFT_BL_ON;
// init TFT library
  Tft.TFTinit();
  Tft.fillRectangle(0, 0, screenWidth, screenHeight, BLACK);
  displayButtons();
}

void loop()
{
    // a point object holds x y and z coordinates.
    Point p = ts.getPoint();
    p.x = map(p.x, TS_MINX, TS_MAXX, 0, 240);
    p.y = map(p.y, TS_MINY, TS_MAXY, 0, 320);

    // we have some minimum pressure we consider 'valid'
    // pressure of 0 means no pressing!
    if (p.z == 0 && startTouch == 0) {
      if (selection > -1){
        if (selection < 9 || selection == 10) {
        combined = combined + buttonText[selection];
        lenCombined = lenCombined + 1;
        char toDisplay[lenCombined+1];
        combined.toCharArray(toDisplay,lenCombined+1);
        Tft.drawString(toDisplay, 0, 0, textSize, WHITE);
        }
        if (selection == 11) {
        lenCombined = lenCombined - 1;
        combined = combined.substring(0,lenCombined);
        char toDisplay[lenCombined + 1];
        combined.toCharArray(toDisplay,lenCombined + 1);
         // Erase previous string
        Tft.fillRectangle(0, 0, screenWidth, textHeight, BLACK);
        Tft.drawString(toDisplay, 0, 0, textSize,WHITE);
```

```
      }
    if (selection == 9) {
      char toDisplay[lenCombined + 1];
      combined.toCharArray(toDisplay,lenCombined + 1);
      value = atol(toDisplay);
       Serial.print ("combined = ");
       Serial.println (value);
      }
      selection = -1;
    }
    startTouch = -1;
    delay (100);
    } // Set to something it will never actually be if screen touched
    if (p.z > MinimumPressure && p.z<10000) {
      if (startTouch == -1) {
      startTouch = 0;
      delay (100); // Prevent bounce. may vary time
      }
      px = p.x;
      py = p.y;
      for (x=0;x<3;x++) {
        for (y=0;y<4;y++) {
          if (px > x * buttonWidth + (x + 1) * buttonHSpace && px < x * buttonWidth + (x + 1) * buttonHSpace + buttonWidth){
            if (py >  y * buttonHeight + (y + 1) * buttonVSpace + topSpace && py <  y * buttonHeight + (y + 1) * buttonVSpace + topSpace + buttonHeight) {
              selection = 3 * y + x;
            }
          }
        }
      }
     delay (100); // Prevent bounce. may vary time
    }
}

void displayButtons() {
  for (x=0;x<3;x++) {
    for (y=0;y<4;y++) {
      Tft.fillRectangle(x * buttonWidth + (x + 1) * buttonHSpace, y *
```

buttonHeight + (y + 1) * buttonVSpace + topSpace, buttonWidth, buttonHeight,CYAN);
    Tft.drawString(buttonText[3 * y + x],x * buttonWidth + (x + 1) * buttonHSpace + buttonWidth/3, y * buttonHeight + (y + 1) * buttonVSpace + topSpace + buttonHeight*.15, textSize,textColor);
  }
 }
}

Sketch 7.1

At the beginning of the sketch, a few new variables are set to allow you to easily change the appearance of the numeric keyboard if you like. As the names suggest, buttonWidth is the width of each button, buttonHeight is how tall each button is, buttonHSpace is the horizontal space between buttons, buttonVSpace is the vertical spacing between buttons, topSpace is the amount of space above the keypad on the screen. The buttonText string array holds the characters that will be displayed on the buttons.

The displayButtons subroutine displays the buttons on the screen. The fillRectangle statement draws the individual rectangles for each button and the drawstring statement puts the character in each button. The
for (x=0; x<3; x++)
statement cycles through the 3 columns of buttons and the
for (y=0; y<4; y++)
statement cycles through the 4 rows. Notice that the positions and size of the buttons are controlled by the variables discussed above, so you can easily change the characteristics of all the buttons at once.

The statement
if (p.z > MinimumPressure && p.z<10000)
in the loop subroutine is the start of the section that is activated when you press on the screen. First, the sketch tests to see if startTouch = -1, indicating that you have not been previously touching the screen. If so, it sets startTouch to 0 and does a delay to prevent switch bounce. Then, the sketch determines which button is being touched. The
    for (x=0;x<3;x++) {
      for (y=0;y<4;y++) {
statements cycled through each column and row position, thus

checking each button. The next two statements check to see if the x position is below the top of the button and above the bottom and the y position is to the right of the left edge of the button and the left of the right edge of the button. When they find a button that your touch is within this space, it sets the selection variable to that button with the statement

selection = 3 * y + x;

Notice that the positions of the edges of the buttons are again measured using those variables set at the beginning of the program, so if you decide to move or resize the buttons, this automatically changes.

Just like the text menu sketch, when you release the pressure on the screen, as determined by the

if (p.z == 0 && startTouch == 0)

statement, the sketch acts on your selection. First, it determines that you were actually touching a button with the

if (selection > -1)

statement. If selection is still -1, you touched and released the screen at a point that was not within one of the buttons. If selection is less than 9 or equals 10, you pressed one of the number buttons. In that case, the number string stored in buttonText array corresponding to selection is appended onto the string variable combined. For example, if combined is empty and you press "1", combined becomes "1". If you then press "2", combined becomes "12". The variable lenCombined, which stores the length of the string lenCombined, is increased by 1. The statements

char toDisplay[lenCombined+1];
combined.toCharArray(toDisplay,lenCombined+1);
Tft.drawString(toDisplay,0, 0, textSize, WHITE);

print the variable combined at the top of the screen. This may seem a bit complicated, but as explained in Chapter 3, the function drawString has trouble handling the type of string variable that can allow the type of manipulation needed to add the strings together.

If selection is 11, you pressed the "C" key, which stands for clear. This cuts off the last number in the combined string. (You can change this to "B" for backspace or some other character if you prefer simply by changing it when you fill the buttonText array at the beginning of the program.) First, lenCombined is decreased by 1, then the statement

combined = combined.substring(0,lenCombined);

changes combined to the new shorter string. The sketch then uses the same code to print the new shorter string to the display as it uses to print the string when you add a character, except that it contains the statement
Tft.fillRectangle(0, 0, screenWidth, textHeight, BLACK);
to erase the previous string. This was not necessary before, because when it printed the string the new string was simply printed on top of an identical string, except for the last character, which was added. However, when you take off a character, you need to erase the old string or the last character will still be there.

If selection equals 9, you pressed the E key, for Enter. This causes the sketch to actually act on the number you have typed. The statements
char toDisplay[lenCombined + 1];
combined.toCharArray(toDisplay,lenCombined + 1);
value = atol(toDisplay);
convert the string to a long integer, value. The variable value could also be a regular integer if you are sure nobody will input a value beyond the maximum integer value, 32767.

The result of this sketch is shown in Figure 7.1, with some random numbers typed in.

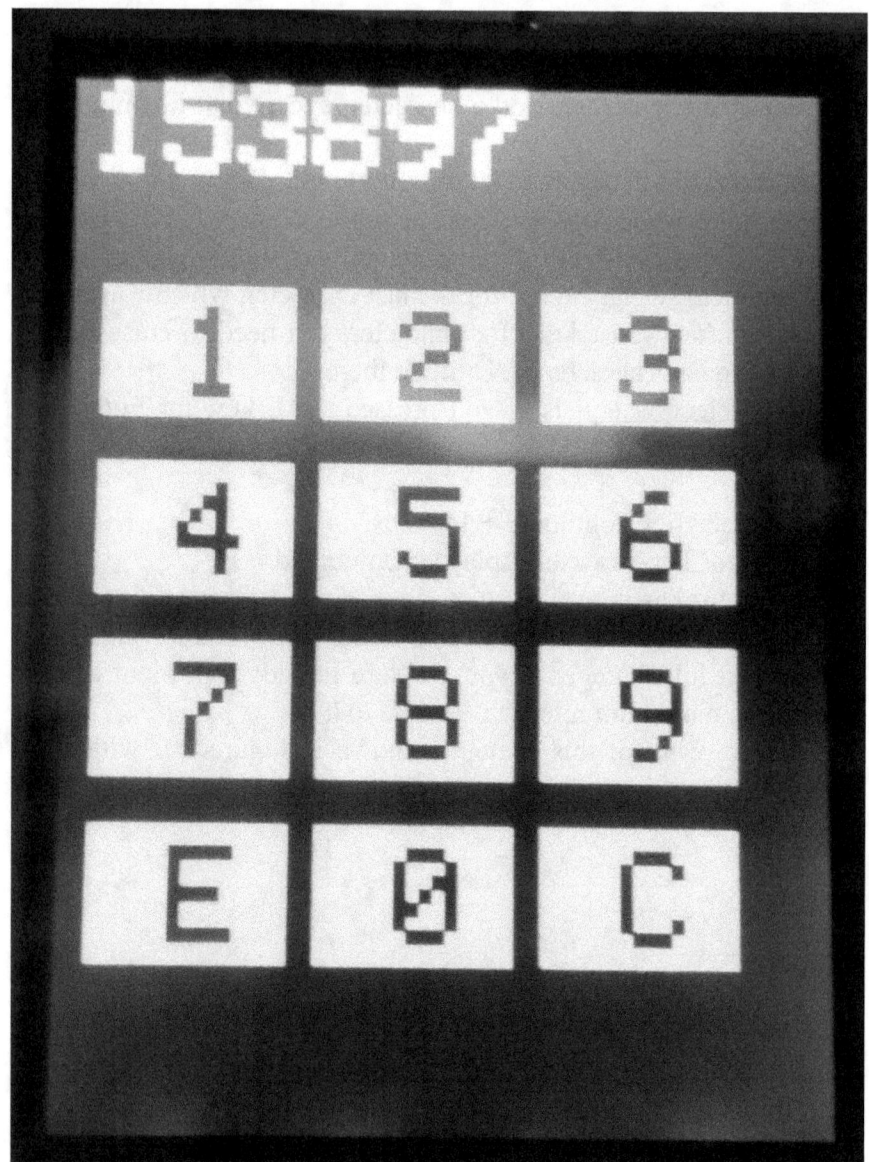

Figure 7.1

In this sample program, that value is simply sent to the serial port. However, in your actual program, you can use it to do something more interesting. For example, you could use it to set the duty cycle of a pulse width modulated digital output pin. You would do this by replacing
Serial.print ("combined = ");
Serial.println (value);

with the command
analogWrite(pin,value);
where pin is a PWM digital capable pin you have previously defined as an output pin. The allowable range for this is 0 to 255. This would allow you to control the brightness of a light, the speed of a motor, or other functions. Of course, you could add simple code like value=value/100*255 to allow you to input a simpler number like 0 to 100 and have it convert that value to the range of the PWM pin.

    Note: If you want to delete the value of the numbers after you have performed whatever function you use for the execute command, you can add
combined="";
lenCombined = 0;
Tft.fillRectangle(0, 0, screenWidth, textHeight, BLACK);
after the command such as analogWrite or Serial.println. This will set the string to empty and clear the number from the display, so you can then type another number. Otherwise, you would have to hit the "C" key until the number is gone. On the other hand, if you want a visual record of what number was last entered, you can leave it there.

    Notice that the key is not acted on until you take your finger off the screen, like in the text menu. This will allow you to slide your finger on the screen to a different button is you see that you are not touching the button you want to. Like with the text menu, you can add a visual indicator of what button you are pressing. This requires a little tweaking of the displayButtons subroutine. Before the Tft.fillRectangle statement, insert the lines
buttonColor=CYAN;
if (selection==3 * y + x) {buttonColor=RED;}
and in the Tft.fillRectangle statement, change the CYAN to buttonColor. Then, add displayButtons(); after the statement
selection = 3 * y + x;
and after the
selection = -1;
statement in the code that acts on whatever button has been pressed.

    Once you make these changes, the display will change the color of the button your touch to red as you press down on it and change it back to cyan when you release it. If you find that the wrong button has changed color, you can slide your finger to the correct button before you take your finger off the screen. The buttons will change color as you move over them, but the selection will not be made until you take your finger off the screen. Sketch 7.2 shows the complete code with the color changing buttons and the erasure of the numbers at the top of the screen when you press the E button.

```
#include <TFTv2.h>
#include <SPI.h>
#include <SeeedTouchScreen.h>

int screenHeight = 320;
int screenWidth = 240;
int MinimumPressure = 50; //Minimum pressure to consider touch

int textSize = 4;
int textColor = BLACK;
int buttonWidth = 60;
int buttonHeight = 40;
int buttonHSpace = 10;
int buttonVSpace = 15;
int topSpace = 50;
char* buttonText[] =
{"1","2","3","4","5","6","7","8","9","E","0","C"};

int x;
int y;
int px;
int py;
int selection = -1;
int startTouch = -1;
String combined = "";
int lenCombined = 0;
long value;
int buttonColor;
int textHeight = 7.5 * textSize;

TouchScreen ts = TouchScreen(XP, YP, XM, YM); //init
TouchScreen port pins

void setup()
{
```

```
  Serial.begin(9600);

  TFT_BL_ON;
// init TFT library
  Tft.TFTinit();
  Tft.fillRectangle(0, 0, screenWidth, screenHeight, BLACK);
  displayButtons();
}

void loop()
{
    // a point object holds x y and z coordinates.
   Point p = ts.getPoint();
   p.x = map(p.x, TS_MINX, TS_MAXX, 0, 240);
   p.y = map(p.y, TS_MINY, TS_MAXY, 0, 320);

   // we have some minimum pressure we consider 'valid'
   // pressure of 0 means no pressing!
   if (p.z == 0 && startTouch == 0) {
    if (selection > -1){
      if (selection < 9 || selection == 10) {
      combined = combined + buttonText[selection];
      lenCombined = lenCombined + 1;
      char toDisplay[lenCombined+1];
      combined.toCharArray(toDisplay,lenCombined+1);
      Tft.drawString(toDisplay,0, 0, textSize,WHITE);
      }
      if (selection == 11) {
      lenCombined = lenCombined - 1;
      combined = combined.substring(0,lenCombined);
      char toDisplay[lenCombined + 1];
      combined.toCharArray(toDisplay,lenCombined + 1);
       // Erase previous string
      Tft.fillRectangle(0, 0, screenWidth, textHeight, BLACK);
      Tft.drawString(toDisplay, 0, 0, textSize,WHITE);
      }
      if (selection == 9) {
      char toDisplay[lenCombined + 1];
      combined.toCharArray(toDisplay,lenCombined + 1);
      value = atol(toDisplay);
```

```
      Serial.print ("combined = ");
      Serial.println (value);
      combined="";
      lenCombined = 0;
      Tft.fillRectangle(0, 0, screenWidth, textHeight, BLACK);
    }
    selection = -1;
    displayButtons();
  }
  startTouch = -1;
  delay (100);
  } // Set to something it will never actually be if screen touched
  if (p.z > MinimumPressure && p.z<10000) {
    if (startTouch == -1) {
      startTouch = 0;
      delay (100); // Prevent bounce. may vary time
    }
    px = p.x;
    py = p.y;
    for (x=0;x<3;x++) {
      for (y=0;y<4;y++) {
        if (px > x * buttonWidth + (x + 1) * buttonHSpace && px < x * buttonWidth + (x + 1) * buttonHSpace + buttonWidth){
          if (py >  y * buttonHeight + (y + 1) * buttonVSpace + topSpace && py <  y * buttonHeight + (y + 1) * buttonVSpace + topSpace + buttonHeight) {
            selection = 3 * y + x;
            displayButtons();
          }
        }
      }
    }
    delay (100); // Prevent bounce. may vary time
  }
}

void displayButtons() {
  for (x=0;x<3;x++) {
    for (y=0;y<4;y++) {
      buttonColor=CYAN;
```

```
    if (selection==3 * y + x) {buttonColor=RED;}
   Tft.fillRectangle(x * buttonWidth + (x + 1) * buttonHSpace, y *
buttonHeight + (y + 1) * buttonVSpace + topSpace, buttonWidth,
buttonHeight,buttonColor);
   Tft.drawString(buttonText[3 * y + x],x * buttonWidth + (x + 1) *
buttonHSpace + buttonWidth/3, y * buttonHeight + (y + 1) *
buttonVSpace + topSpace + buttonHeight*.15, textSize,textColor);
  }
 }
}
```

Sketch 7.2

Chapter 8

Buttons

We discussed numeric keypads in the previous chapter. Now let's get into more generalized buttons. Sketch 8.1 shows how to put an array of buttons on the screen and respond when you press and release one. I have designed this to be very easily altered to change the button sizes, positions, and other characteristics.

```
#include <TFTv2.h>
#include <SPI.h>
#include <SeeedTouchScreen.h>
int screenHeight = 320;
int screenWidth = 240;
int MinimumPressure = 50;
int textSize = 3;
int textHeight = 7.5 * textSize;
int textWidth = 6 * textSize;
int stringLength;
String S;
int x;
int y;
int px;
int py;
int selection = -1;
int previousSelection = -1;
int startTouch = -1;
int buttonColor;
int textColor = BLACK;
int normalButtonColor=CYAN;
int activeButtonColor=RED;
int buttonWidth[] = {200,220,100,60,120};
int buttonHeight[] = {40,40,40,50,40};
int buttonTop[] = {10, 70,125,190,260};
int buttonLeft[] = {20,10,60,90,60};
char* buttonText[] = {"Activate", "Deactivate", "Stop", "Go", "Pause"};
```

```
int numButtons = 5;

TouchScreen ts = TouchScreen(XP, YP, XM, YM);

void setup()
{
  Serial.begin(9600);
  TFT_BL_ON;
  Tft.TFTinit();
  Tft.fillRectangle(0, 0, screenWidth, screenHeight, BLACK);
  displayButtons();
}

void loop()
{
    // a point object holds x y and z coordinates.
    Point p = ts.getPoint();
    p.x = map(p.x, TS_MINX, TS_MAXX, 0, 240);
    p.y = map(p.y, TS_MINY, TS_MAXY, 0, 320);

    if (p.z == 0 && startTouch == 0) {
      if (selection > -1){
        if (selection == 0) {
          Serial.println ("First");
        }
        if (selection == 1) {
          Serial.println ("Second");
        }
        if (selection == 2) {
          Serial.println ("Third");
        }
        if (selection == 3) {
          Serial.println ("Fourth");
        }
        if (selection == 4) {
          Serial.println ("Fifth");
        }
        selection = -1;
        previousSelection = -1;
        displayButtons();
```

```
      }
    startTouch = -1;
    delay (100);
    }
  if (p.z > MinimumPressure && p.z<10000) {
    if (startTouch == -1) {
      startTouch = 0;
      delay (100);
      }
    px = p.x;
    py = p.y;
    selection = -1;
    for (x=0;x<numButtons;x++) {
         if (py > buttonTop[x] && py < buttonTop[x] +
buttonHeight[x]) {
            if (px > buttonLeft[x] && px < buttonLeft[x] +
buttonWidth[x]) {
              selection =  x;
             }
           }
         }
      if (previousSelection != selection){
         displayButtons();
         }
      previousSelection = selection;
      delay (100);
    }
}

void displayButtons() {
  for (x=0;x<numButtons;x++) {
    S = buttonText[x];
    stringLength = S.length();
    buttonColor = normalButtonColor;
    if (selection == x) {buttonColor = activeButtonColor;}
    Tft.fillRectangle(buttonLeft[x], buttonTop[x], buttonWidth[x],
buttonHeight[x],buttonColor);
    Tft.drawString(buttonText[x], buttonLeft[x] + (buttonWidth[x] -
(stringLength * textWidth))/2, buttonTop[x] + (buttonHeight[x]-
textHeight)/2 , textSize, textColor);
```

}
}

Sketch 8.1

This particular sketch displays five buttons, as set by the line
int numButtons = 5;
The array buttonWidth sets the width of each button, buttonHeight sets the height of each button, buttonTop sets the position of the top of the button on the screen, and buttonLeft sets the left of the button on the screen. These are all measured in pixels on the screen. The array buttonText is the text of the buttons. The variable textColor is the color of the text within the buttons. The variable normalButtonColor is the color of the buttons as normally seen on the screen. The variable activeButtonColor s the color the button changes to as you press it to indicate which button you are pressing.

The subroutine displayButtons displays the buttons. The statement
for (x=0;x<numButtons;x++)
counts through the buttons. The statement
S = buttonText[x];
saves the text in the character array element buttonText[x] into the string variable S. The statement
stringLength = S.length();
then determines the length of the text. This is used to center the text within the button. The statements
buttonColor = normalButtonColor;
if (selection == x) {buttonColor = activeButtonColor;}
set the button color to its normal color, but then change that color to activeButtonColor if that button is being pressed. The statement

Tft.fillRectangle(buttonLeft[x], buttonTop[x], buttonWidth[x], buttonHeight[x],buttonColor);

draws the button, and then

Tft.drawString(buttonText[x],buttonLeft[x] + (buttonWidth[x]-(stringLength * textWidth))/2, buttonTop[x] + (buttonHeight[x]-textHeight)/2 , textSize, textColor);

draws the text within the button, centered both horizontally and vertically within the button. One minor note: If you want to center each button, you can dispense with the buttonLeft array entirely and replace buttonLeft[x] throughout the code with (screenWidth - buttonWidth[x])/2).

Within the loop, the code
```
if (p.z > MinimumPressure && p.z<10000)
```
checks to see if the screen is being touched. If so, it sets startTouch to 0 to indicate that the screen has been touched. Then the code
```
px = p.x;
py = p.y;
selection = -1;
for (x=0;x<numButtons;x++) {
   if (py > buttonTop[x] && py < buttonTop[x] + buttonHeight[x]) {
     if (px > buttonLeft[x] && px < buttonLeft[x] +
     buttonWidth[x]) {
        selection = x;
      }
    }
  }
  if (previousSelection != selection){
    displayButtons();
  }
  previousSelection = selection;
```
determines which button is being pressed. The variables px and py merely store the touch position for easy reference. The for loop cycles through each button in the array. The if statements check to see if the touch is right of the left edge of the button and left of the right edge and below the top and above the bottom. If all these conditions are met, the value of selection is set to that value of x. Note that the button arrays start with 0, so the first button is 0, not 1. Note also that selection is set to -1 before the for loop, so even if you has previously been pressing a button and selection was set to that value, if you slide your finger off the button and the for loop does not find a button you are pressing, the value of selection will then indicate that no button is being touched. This allows you to cancel any button press without selecting another button just by moving your finger to a blank place on the screen.

Once selection is determined, the sketch calls the displayButtons subroutine again. This is so the selected button can be shown in a different color to show you which button you are pressing. However, in order to prevent the screen from flickering every time the Arduino gets to this part of the code, it only does this if the selection has changed while you are still touching the screen. The variable previousSelection records the latest selection. The statement

if (previousSelection != selection)

then compares the selection just found with this value, and only calls the display if they are not equal. The value of previousSelection is set after this test.

Once you have pressed and then released a button, the code within the

if (p.z == 0 && startTouch == 0)

test takes effect. If p.z equals 0 (indicating that you are not touching the screen) and startTouch equals 0 (indicating that you had previously been touching the screen), the if statement is satisfied and the code within it is executed. In this case, it first tests to make sure selection is not -1, which would indicate that you were not touching any button when you took your finger off the screen. It then goes through an if test for each possible value of selection to determine what to do for that selection. In this example program, it simply sends a message to the serial output. Of course, in your actual sketch, you would have it do something like send a digital output HIGH or LOW, or some other significant action.

The resulting screen is shown in Figure 8.1.

Figure 8.1

Chapter 9

Histograms and Sliders

In this chapter, we will discuss histograms, an interesting way to display data on a graphic screen. We will go beyond displaying data with a histogram, however, and show how to use a histogram to set the value of a variable, in the manner of a slider control.

First, let's look at simple histograms. These do not actually use the touch feature of the touch screen, because they are only displaying data. However, they can be useful for displaying information, such as the values of the analog inputs. What is displayed on the screen is a rectangle that is part cyan in color and part red. If the value being displayed is 0, the entire rectangle is cyan. If the value being displayed is 100% of its maximum value, the rectangle is entirely red. If the value is 50% of its maximum value, the rectangle is 50% cyan and 50% red, and so on. The cyan portion is referred to in this chapter as the portion that is not filled in, and the red portion is referred to as the portion that is filled in.

Although histograms do not display the data as precisely as outputting an actual number, they can be useful for showing rapidly changing data, like values of the analog inputs. You can also have the screen display the actual numeric value, as well as displaying it graphically with the histogram, if you want actual values in addition to the graphic representation of the values. You can have buttons or other controls on the screen along with the histograms, so you can respond to readings or control the readings. Sketch 9.1 shows sketch for displaying such histograms.

```
#include <TFTv2.h>
#include <SPI.h>
#include <SeeedTouchScreen.h>

int screenHeight = 320;
int screenWidth = 240;
int textSize = 2;
int textColor = WHITE;
```

```
int x;
bool readingChanged;
float fractionValue;
int numHistograms = 5;
int histogramColor = CYAN;
int fillColor = RED;
int histogramWidth[] = {30,30,30,30,30};
float histogramHeight[] = {260,260,260,260,260};
float histogramTop[] = {20, 20,20,20,20};
int histogramLeft[] = {5,55,105,155,205};
float histogramValue[] = {0,0,0,0,0};
float histogramMinValue[] = {0,0,0,0,0};
float histogramMaxValue[] = {1023,1023,1023,1023,1023};
int histogramPreviousValue[] = {0,0,0,0,0};
char* textLabel[] = {"A4","A5","A6","A7","A8"};

TouchScreen ts = TouchScreen(XP, YP, XM, YM);

void setup()
{
 Serial.begin(9600);
 TFT_BL_ON;
 Tft.TFTinit();
 Tft.fillRectangle(0, 0, screenWidth, screenHeight, BLACK);
 displayHistograms();
}

void loop(){
  readingChanged = false;
  for (x=0;x<numHistograms;x++) {
   histogramValue[x] = analogRead(x + 4);
   if (histogramValue[x] != histogramPreviousValue[x]){
    readingChanged = true;
    histogramPreviousValue[x] = histogramValue[x];
   }
  }
  if (readingChanged){
   displayHistograms();
   }
}
```

```
void displayHistograms() {
  for (x=0;x<numHistograms;x++) {
    fractionValue = (histogramValue[x] - histogramMinValue[x])/
(histogramMaxValue[x] - histogramMinValue[x]);
    if (fractionValue < 1) {
      Tft.fillRectangle(histogramLeft[x], histogramTop[x],
histogramWidth[x], histogramHeight[x] * (1 - fractionValue)-1,
histogramColor);
    }
    if (fractionValue > 0){
      Tft.fillRectangle(histogramLeft[x], histogramTop[x] + (1-
fractionValue) * histogramHeight[x], histogramWidth[x],
histogramHeight[x] * fractionValue, fillColor);
    }
    Tft.drawString(textLabel[x], histogramLeft[x], histogramTop[x] +
histogramHeight[x]+3, textSize, textColor);
  }
}
```

Sketch 9.1

First, we have some variables to define. Many are familiar from previous sketches. You might notice that I have made the text size a little smaller, because we will be crowding it in with the histograms.

The Boolean variable readingChanged is used to signal when a value being displayed by the histograms has changed, and hence the display needs to be updated. The variable fractionValue is a number from 0 to 1 that tells how much of the rectangle to fill, where 1 would fill it completely and represent that the value being displayed is 100% of its maximum possible value. The variable numHistograms is the number of histograms to display on the screen, histogramColor is the background color of the histogram, and fillColor is the color of the part that is filled in.

These variables are followed by quite a few arrays, to allow you to have any number of histograms and set values for the properties of each one. The histogramWidth[] is the width of a histogram on the screen, histogramHeight[] is the height and histogramTop[] is the top position of each histogram on the screen.

Although I have used arrays for these, in practice you will probably want all histograms to be the same height and width and top, so you could actually use single numbers for these instead of arrays. That is,
int histogramWidth[] = {30,30,30,30,30};
float histogramHeight[] = {260,260,260,260,260};
float histogramTop[] = {20, 20,20,20,20};
could be replaced by just
int histogramWidth = 30;
float histogramHeight = 260;
float histogramTop = 20;
and you would just use histogramWidth, histogramHeight, and histogramTop throughout the sketch instead of histogramWidth[x], histogramHeight[x], and histogramTop[x]. However, I have written the sketch using arrays for these to allow you for flexibility if you want it.

    The variable histogramLeft[] gives the position of the left of each histogram, and does need to be different from each histogram or they would all be in the same place. The variables histogramValue[], histogramMinValue[], and histogramMaxValue[] store the current value of the number the histogram is portraying, the minimum value it will display, and the maximum value it will display, respectively. I have dimensioned histogramValue[] with zeros just to start the program. In this example, I have set the maximum and minimum values to 1023 and 0, the possible range of the analog inputs.

    You can change the number of histograms within the program, allowing you to display different numbers of histograms at different times. If you do plan on changing numHistograms within the program, be sure to dimension histogramValue[] with the maximum value that you ever plan to use for numHistograms. For that matter, you will need to dimension all of the arrays with the maximum number you will ever use by providing enough numbers in the lines that dimension them. The variable histogramPreviousValue[] is the value of histogramValue[] the previous time through the loop, and is used to determine if the value has changed. The variable textLabel[] stores the labels that will be displayed under the histograms so you know what each one is.

    Regarding the values of histogramMinValue[], it is generally best to set these to 0. However, if you do set it to another number, be sure that the value of histogramValue[] is never less than histogramMinValue[] for the corresponding element. If it is, the bottom of the histogram will suddenly go down below where it should be on the screen, possibly even below the end of the screen. While this does no harm, it will make the display look messy. If you do set any values of histogramMinValue[] more than 0, you must set the corresponding initial values of histogramValue[] and histogramPreviousValue[] to this minimum value instead of 0 for the same reason.

Let's jump now to the displayHistograms subroutine. The for loop for (x=0;x<numHistograms;x++) goes through all the histograms to be displayed. As explained previously, you can change the value of numHistograms at any time and the subroutine will display only that many histograms. For each histogram, the first line within the for loop determines the fraction of the histogram to fill, as explained previously. If the minimum value of all histograms is 0, as I recommend, you can simplify this to
fractionValue = histogramValue[x] /histogramMaxValue[x];
and eliminate the line at the beginning of the program that dimensions the histogramMinValue array.

The first Tft.fillRectangle command, inside the if (fractionValue < 1) condition, draws the cyan portion of the histogram. The reason that it is within an if condition is that even if the value of fractionValue is 1, in which case there should be no cyan showing, the Tft.fillRectangle command draws a slight line. The if condition prevents it from drawing anything if the value is 1.

The second Tft.fillRectangle command, inside the if (fractionValue > 0) condition, draws the red portion of the rectangle. The reason that it is within an if condition is that even if the value of fractionValue is 0, in which case there should be no red showing, the Tft.fillRectangle command draws a slight line. The if condition prevents it from drawing anything if the value is 0.

Within the main loop function, we first set the flag readingChanged to false. Then the for loop goes through each histogram. It first sets the value of the histogramValue array variable for element x. In this example, it reads this value from an analog port. The if statement then determines if the value is different from the previous value. If so, it sets the readingChanged flag to true and also sets the histogramPreviousValue[x] to the current value, updating this test. Once it has set the value for each histogram and tested to see if it has changed, it checks to see if readingChanged has been set to true. If so, it calls the displayHistograms subroutine to display the new readings.

Of course, we could have simply had the code call the displayHistograms subroutine as soon as it first found a change instead of setting a flag and then testing the flag later. (This would, incidentally, have required that we use a different variable for the for loop in the displayHistograms, or else defining x locally instead of globally.) However, this would have called the displayHistograms subroutine for each of the analog inputs (if they were all changing). This would make the program run much more slowly, because the displayHistograms routine would redraw all of the histograms each time one changed. It would make the display look choppy, because you would see each histogram change separately. Setting a flag and then changing al the histograms at once is much better.

One detail you might have noticed is that the histogram labels read A4 through A8, and the analogRead function was reading x+4 so that it was reading A4 through A8 when x went from 0 to 4. This is because the touch screen from Seeed Studio uses the analog pins A0 through A3, so you cannot read those. This code would therefore require a more advanced board, like an Arduino Mega. Of course, if you had fewer inputs it would run on an Arduino. Figure 9.1 shows the screen.

Figure 9.1

Early in this chapter, I mentioned the possibility of putting other controls on the screen along with the histograms. My main point was to clarify that the histograms do not have to consume the entire screen. To demonstrate this point, Sketch 9.2, combines the buttons program from Chapter 8 (Sketch 8.1) with Sketch 9.1.

#include <TFTv2.h>

```cpp
#include <SPI.h>
#include <SeeedTouchScreen.h>

int screenHeight = 320;
int screenWidth = 240;
int MinimumPressure = 50;
int textSize = 3;
int textHeight = 7.5 * textSize;
int textWidth = 6 * textSize;
int stringLength;
String S;
int v[2];
int y;
int px;
int py;
int selection = -1;
int previousSelection = -1;
int startTouch = -1;
int buttonColor;
int textColor = WHITE;
int normalButtonColor=GREEN;
int activeButtonColor=RED;
int buttonWidth[] = {90,90,90,90};
int buttonHeight[] = {40,40,40,40};
int buttonTop[] = {20,90,160,220};
int buttonLeft[] = {150,150,150,150};
char* buttonText[] = {"V1+", "V1-", "V2+", "V2-"};
int numButtons = 4;
bool readingChanged;
float fractionValue;
int numHistograms = 2;
int histogramColor = CYAN;
int fillColor = RED;
int histogramWidth[] = {30,30};
float histogramHeight[] = {260,260};
float histogramTop[] = {20, 20};
int histogramLeft[] = {5,55};
float histogramValue[] = {0,0};
float histogramMinValue[] = {0,0};
float histogramMaxValue[] = {1023,1023};
```

```
int histogramPreviousValue[] = {0,0};
char* textLabel[] = {"V1","V2"};
TouchScreen ts = TouchScreen(XP, YP, XM, YM);

void setup()
{
 TFT_BL_ON;
 Tft.TFTinit();
 Tft.fillRectangle(0, 0, screenWidth, screenHeight, BLACK);
 displayButtons();
 displayHistograms();
}

void loop(){
   int x;
    // a point object holds x y and z coordinates.
   Point p = ts.getPoint();
   p.x = map(p.x, TS_MINX, TS_MAXX, 0, 240);
   p.y = map(p.y, TS_MINY, TS_MAXY, 0, 320);

   if (p.z == 0 && startTouch == 0) {
    if (selection > -1){
      if (selection == 0) {
       v[0] = v[0] + 100;
        if (v[0] > histogramMaxValue[0]) {v[0] = histogramMaxValue[0];}
      }
      if (selection == 1) {
       v[0] = v[0] - 100;
        if (v[0] < histogramMinValue[0]) {v[0] = histogramMinValue[0];}
      }
      if (selection == 2) {
       v[1] = v[1] + 100;
        if (v[1] > histogramMaxValue[1]) {v[1] = histogramMaxValue[1];}
      }
      if (selection == 3) {
       v[1] = v[1] - 100;
        if (v[1] < histogramMinValue[1]) {v[1] =
```

```
      histogramMinValue[1];}
        }
        selection = -1;
        previousSelection = -1;
        displayButtons();
      }
      startTouch = -1;
      delay (100);
      }
    if (p.z > MinimumPressure && p.z<10000) {
      if (startTouch == -1) {
        startTouch = 0;
        delay (100);
        }
        px = p.x;
        py = p.y;
        selection = -1;
        for (x=0;x<numButtons;x++) {
          if (py > buttonTop[x] && py < buttonTop[x] + buttonHeight[x]) {
            if (px > buttonLeft[x] && px < buttonLeft[x] + buttonWidth[x]) {
              selection = x;
            }
          }
        }
        if (previousSelection != selection){
          displayButtons();
        }
        previousSelection = selection;
      delay (100);
      }
    readingChanged = false;
    for (x=0;x<numHistograms;x++) {
     histogramValue[x] = v[x];
     if (histogramValue[x] != histogramPreviousValue[x]){
       readingChanged = true;
       histogramPreviousValue[x] = histogramValue[x];
     }
    }
```

```
  if (readingChanged){
    displayHistograms();
    }
}

void displayButtons() {
  int x;
  for (x=0;x<numButtons;x++) {
    S = buttonText[x];
    stringLength = S.length();
    buttonColor = normalButtonColor;
    if (selection == x) {buttonColor = activeButtonColor;}
    Tft.fillRectangle(buttonLeft[x], buttonTop[x], buttonWidth[x], buttonHeight[x],buttonColor);
    Tft.drawString(buttonText[x],buttonLeft[x] + (buttonWidth[x]-(stringLength * textWidth))/2, buttonTop[x] + (buttonHeight[x]-textHeight)/2 , textSize,textColor);
  }
}


void displayHistograms() {
  int x;
  for (x=0;x<numHistograms;x++) {
    fractionValue = (histogramValue[x] - histogramMinValue[x])/(histogramMaxValue[x] - histogramMinValue[x]);
    if (fractionValue < 1) {
      Tft.fillRectangle(histogramLeft[x], histogramTop[x], histogramWidth[x], histogramHeight[x] * (1 - fractionValue)-1, histogramColor);
    }
    if (fractionValue > 0){
      Tft.fillRectangle(histogramLeft[x], histogramTop[x]+(1-fractionValue) * histogramHeight[x], histogramWidth[x], histogramHeight[x] * fractionValue,fillColor);
    }
    Tft.drawString(textLabel[x], histogramLeft[x], histogramTop[x] + histogramHeight[x] + 3, textSize, textColor);
  }
}
```

Sketch 9.2

This sketch puts two histograms and four buttons on the screen. In this example, the buttons actually control the value the variables (stored in the v[] array) that are displayed by the histograms. The histograms are simply labeled V1 and V2 (for variable 1 and 2) and the buttons are labeled V1+, V1-, V2+, and V2- to indicate that the buttons increase or decrease the value of the variables V1 and V2.

The sketch contains the subroutines displayHistograms from Sketch 9.1 and displayButtons from Sketch 8.1 with only one small change. That is that the counting variable x is defined locally within the routine. This is so that counting with this variable within these subroutines does not alter the x count within the routine that called the subroutine. This ideal of locally verses globally defined variables can be useful in coding for the Arduino to prevent such unwanted changes.

Sketch 9.2 also contains the variables from the beginning of each program. Some of the preset values are different in value or number of dimensions, because we have changed the number of histograms and buttons from the original sketches. For example, textLabel[] is now assigned only two values V1 and V2. The setup routine runs both displayButtons and displayHistograms to get you started.

The main difference is in the loop routine where it handles the response to pressing and then releasing buttons. In Sketch 8.1, the code simply sent a message such as "First" or "Second" to the serial port to indicate which button had been pressed. In this one, the code changes the value of the variables v[0] and v[1] based on which button has been pressed and released. For example, if the button labeled 1+, which is button 0, is pressed and released the value of v[0] increases by 100. If the button labeled 1-, which is button 1, is pressed and released, v[0] deceases by 100. Notice that there is a safeguard in the code against the values increasing or decreasing above or below their limits. For example, after the code that increases v[0] by 100, you have the code
if (v[0] > histogramMaxValue[0]) {v[0] = histogramMaxValue[0];}
This says that if v[0] has gone above the limit for the histogram, reset it to the limit.

At the end of the loop routine, we have code very similar to the code in Sketch 9.1 that checks to see if the values have changed and therefore it is necessary to redraw the histograms. The only difference is that instead of
histogramValue[x] = analogRead(x+4);
where the Arduino reads the analog port, we have
histogramValue[x] = v[x];
where the histogram value is set to the variable we have been changing with the buttons.

Note that x is defined at the beginning of the loop routine. It has to be, since it is not defined globally to prevent conflict.

The result of this sketch is shown in figure 9.2.

Figure 9.2

In Sketch 8.1 and in Sketch 9.2, we deliberately made it so that you had to press and release the button to trigger the action from the button. In most cases, you want that to prevent repeat triggering when you hold down the button. However, in some cases you want that. This is a good example. Since pressing the button creates a change in the value, you might want to have it simply holding down the button causes a continuous increase or decrease in the value as long as you hold down the button. Doing this is very easy. Just

remove (or remark out) a few lines of the code. These lines are
if (p.z == 0 && startTouch == 0) {
and the corresponding }, which is the line directly after the first delay(100) and directly above the line
if (p.z > MinimumPressure && p.z<10000) {
You probably will also want to remove or decrease some or all of the delays. The drawing of the histograms actually takes a fraction of a second anyway, so the delays are unnecessary if you want continuous changes while you hold down the buttons.

    The above examples demonstrate uses for buttons, but there is a more fun and efficient way to change the values of the histograms. We can write the code so that just touching a histogram changes the value of that histogram to the value where you touched it. For example, touching a histogram half way up will change the value to 50%. You can actually use the histogram to set values, not just read them. Of course, setting values this way is not quite as precise as actually inputting numbers. However, for many purposes, such as setting light or sound levels, it is quite accurate enough. Sketch 9.3 shows how to do this.

```
#include <TFTv2.h>
#include <SPI.h>
#include <SeeedTouchScreen.h>

int screenHeight = 320;
int screenWidth = 240;
int MinimumPressure = 50;
int textSize = 2;
int textHeight = 7.5 * textSize;
int textWidth = 6 * textSize;
int textColor = WHITE;

int x;
int px;
int py;
int selection = -1;
int previousSelection = -1;
int startTouch = -1;
float fractionValue;
```

```
int fillColor = RED;
int sliderColor = CYAN;
int sliderWidth[] = {30,30,30,30,30};
float sliderHeight[] = {260,260,260,260,260};
float sliderTop[] = {20, 20,20,20,20};
int sliderLeft[] = {5,55,105,155,205};
float sliderValue[] = {110,25,50,75,100};
float sliderMinValue[] = {100,0,0,0,0};
float sliderMaxValue[] = {200,100,100,100,100};

char* textLabel[] = {"A","B","C","D","E"};
int numSliders = 5;

TouchScreen ts = TouchScreen(XP, YP, XM, YM);

void setup()
{
  Serial.begin(9600);
  TFT_BL_ON;
  Tft.TFTinit();
  Tft.fillRectangle(0, 0, screenWidth, screenHeight, BLACK);
  displaySliders();
}

void loop()
{
   Point p = ts.getPoint();
   p.x = map(p.x, TS_MINX, TS_MAXX, 0, 240);
   p.y = map(p.y, TS_MINY, TS_MAXY, 0, 320);
   if (p.z == 0 && startTouch == 0) {
    if (selection > -1){
      if (selection == 0) {
       Serial.print ("First =");
       Serial.println (sliderValue[selection]);
       }
      if (selection == 1) {
       Serial.print ("Second =");
       Serial.println (sliderValue[selection]);
       }
      if (selection == 2) {
```

```
    Serial.print ("Third =");
    Serial.println (sliderValue[selection]);
   }
   if (selection == 3) {
    Serial.print ("Fourth = ");
    Serial.println (sliderValue[selection]);
   }
   if (selection == 4) {
    Serial.print ("Fifth = ");
    Serial.println (sliderValue[selection]);
   }
   selection = -1;
   previousSelection = -1;
  }
  startTouch = -1;
  delay (100);
 }
 if (p.z > MinimumPressure && p.z<10000) {
   if (startTouch == -1) {
     startTouch = 0;
     delay (100);
    }
    px = p.x;
    py = p.y;
    selection = -1;
    for (x=0;x<numSliders;x++) {
       if (py>sliderTop[x]-20 && py<sliderTop[x] + sliderHeight[x]+20) {
         if (px > sliderLeft[x] && px < sliderLeft[x] + sliderWidth[x]) {
         selection = x;
         sliderValue[x] = int(((sliderTop[x] + sliderHeight[x]) - py) /sliderHeight[x] * (sliderMaxValue[x]-sliderMinValue[x]) + sliderMinValue[x]);
         if (py < sliderTop[x]) {sliderValue[x] = sliderMaxValue[x];}
         if (py > sliderTop[x]+sliderHeight[x]) {sliderValue[x] = sliderMinValue[x];}
         }
       }
```

```
            }
          if (previousSelection != selection){
            displaySliders();
          }
          previousSelection = selection;
      }
  }

  void displaySliders() {
    for (x=0;x<numSliders;x++) {
      fractionValue = (sliderValue[x] - sliderMinValue[x])/(sliderMaxValue[x] - sliderMinValue[x]);
      Tft.fillRectangle(sliderLeft[x], sliderTop[x], sliderWidth[x], sliderHeight[x] * (1 - fractionValue), sliderColor);
      if (fractionValue > 0){
        Tft.fillRectangle(sliderLeft[x], sliderTop[x] + (1-fractionValue) * sliderHeight[x], sliderWidth[x], sliderHeight[x] * fractionValue, fillColor);
      }
      Tft.drawString(textLabel[x], sliderLeft[x], sliderTop[x] + sliderHeight[x] + 3, textSize, textColor);
    }
  }
```

Sketch 9.3

This sketch again combines some features of Sketch 8.1 and 9.1. However, in this case there are no separate buttons. The histograms themselves are the buttons. Touching any slider at any point on the slider and then releasing it will immediately move the value of that slider to the point where you touched it. For example, touching a slider at the top will change the value of that slider to 100%.

In the sketch, one cosmetic change is that the word histogram has been replaced by slider throughout the program. For example, displayHistograms has been replaced by displaySliders. This does not affect the functioning of the program, of course, but it clarifies the functioning when looking at the code. The variables serve the same function, such as sliderWidth setting the width of a slider like histogramWidth set the width of a histogram. The sketch also contains some of the variables used for buttons, such as selection and startTouch. We do not need separate variables to set the position of buttons like buttonTop, since the histograms are the buttons.

Aside from the name change from histogram to slider, the setup routine and the displaySliders routine are the same as the basic histogram program Sketch 9.1. The main loop routine is similar to the loop routine in Sketch 8.1 that read the buttons. For demonstration purposes, I have again simply had the sketch send values to the serial port. However, in this case it is sending actual number values, not simply a notification that a button has been pressed. Looking at the if (p.z == 0 && startTouch == 0) section of the code, you can see that it sends something like "First = 50" to the serial port. Of course, you can change the Serial.print statements to something like analogWrite(controlPin1, sliderValue[selection]) to control the PWM of an output pin. Of course, if you do use the sliders to control PWM, you want to change the sliderMaxValue to 255, the maximum value for an analog output, so that the output can go that high. In Sketch 9.3, I arbitrarily set it to 100 to represent 100%.

The main difference from the plain button code is in the part where you are pressing on a button or slider, the for loop within the if (p.z > MinimumPressure && p.z<10000) condition. In the buttons code, it simply set selection=x if the touch was within the upper, lower, left, and right limits of the button to indicate which button had been pressed. In this code, it still sets the selection variable to x to indicate which slider has been touched, but it also determines where within the slider it has been touched. This is done by the code
sliderValue[x] = int(((sliderTop[x] + sliderHeight[x]) - py) /sliderHeight[x] * (sliderMaxValue[x]-sliderMinValue[x]) + sliderMinValue[x]);
This code determines how high within the slider the touch is as a percentage of the total height of the slider, and then multiplies this by the range of the slider's possible values and adds the minimum value. Note that this allows you to have minimum values other than 0 for the sliders, although in most cases you will want the minimum value to be 0.

Notice in the part of the code that says
if (py>sliderTop[x]-20 && py<sliderTop[x] + sliderHeight[x]+20)
that I have added 20 to the upper and lower ranges where you are allowed to touch the slider and still have it register. This allows you to touch slightly above or below the slider. Although this would give

you a number slightly outside the maximum or minimum value of the slider, the lines

if (py < sliderTop[x]) {sliderValue[x] = sliderMaxValue[x];}

and

if (py > sliderTop[x]+sliderHeight[x]) {sliderValue[x] = sliderMinValue[x];}

reset the value to the maximum or minimum if this happens. This is to allow you to touch the slider slightly beyond the actual slider in order to set the value to minimum or maximum. The reason for this is that I assume that in some cases it might be important to set the values to precisely the minimum (especially 0) or maximum and it can be hard to touch the slider at exactly the top or bottom.

Chapter 10

Plotting Data Like an Oscilloscope

In the previous chapter, we displayed data graphically with histograms. The display of data was transient, with the old values lost as soon as the values changed. It can sometimes be useful to see past and present data plotted over time as a graph. Such a graph being drawn on the screen is sort of a slow motion oscilloscope. The data is plotted in a graph going from left to right on the screen. When the line being drawn reaches the left side of the screen, it starts over on the right side. The old data disappears from each horizontal position when new data is plotted at that position, somewhat like a heart monitor. Sketch 10.1 plots the input from the A4 and A5 analog pins in different colors.

```
#include <SeeedTouchScreen.h>
#include <TFTv2.h>
#include <SPI.h>

const int screenHeight = 320;
const int screenWidth = 240;
const int MinimumPressure = 50;
const int plotBottom = screenHeight-50;
const int plotTop = 10;
const long intervalTime = 100;
const int dotSize = 2;
int minValue = 0;
int maxValue = 1023;
int dataPoint=0;
int dataArray[screenWidth];
int dataArray2[screenWidth];
int x;
int y;
int r;
TouchScreen ts = TouchScreen(XP, YP, XM, YM);

void setup() {
```

```
  Tft.TFTinit();
  TFT_BL_ON;
}

void loop() {
  Tft.fillCircle(dataPoint, dataArray[dataPoint], dotSize, BLACK);
  Tft.fillCircle(dataPoint,dataArray2[dataPoint], dotSize, BLACK);

  r=analogRead(A4);
  y=map(r, minValue, maxValue, plotBottom, plotTop);
  dataArray[dataPoint]=y;
  Tft.fillCircle(dataPoint, y, dotSize, CYAN);

  r=analogRead(A5);
  y=map(r, minValue, maxValue, plotBottom, plotTop);
  dataArray2[dataPoint]=y;
  Tft.fillCircle(dataPoint, y, dotSize, BRIGHT_RED);

  dataPoint++;
  if (dataPoint>=screenWidth){
    dataPoint=0;
  }
  delay(intervalTime);
}
```

Sketch 10.1

First we add some variables and constants to use with the plotter. The variables plotBottom and plotTop define the top and bottom of the graph on the screen. Note that since the y coordinates start at 0 at the top of the screen and go up, the numeric value for the bottom of the screen will be higher than the value for the top of the screen, which can be a little confusing. Setting the top at 0 would have the point plotted right at the top edge of the screen, which could make it hard to see, so a small value like 5 or 10 is best for the top. If it were at 0 and the value plotted happened to stay at maximum, you might not even be sure the points were being plotted. The bottom value can be as high as the screen height minus 1, but again that would not be a good idea for the same reasons. In addition, you might want to put some buttons or text at the bottom of the screen using the techniques described in previous chapters, so setting the bottom of the graph a bit off the bottom of the screen can be useful. The easiest way to do that is to define it in terms of the screen height, like screenHeight - 50.

The variable intervalTime is the time in milliseconds between samples being taken. In this example, I chose 100 milliseconds (1/10th second), but you can set any value you want. For example, a value of 60000 would be one minute. I made intervalTime a long integer variable to leave you lots of options.

The variable dotSize is the size of the dot plotted on the screen. A low value like 1 gives greater precision in the graph, but a higher value like 2 or 3 makes it easier to see the dots. I would recommend a lower number if the line is fairly continuous (values vary little over the time interval), but higher if the values jump around a lot so you can find the points.

The variables minValue and maxValue are the minimum and maximum values you expect to be plotted. Because in this example we are plotting analog input values, I have used 0 to 1023. (Note that some Arduinos have analog inputs that go from 0 to 4095.) Of course, if you expect your possible ranges to be smaller, such as sensor readings that will not vary as much, you can adjust minValue and maxValue. This will cause larger variations in the graph for smaller variations in the readings, making the changes more visible.

The variable dataPoint is the current x coordinate of the plot. This will count up from 0 to screen width - 1. It is set to 0 initially.

The array variables dataArray and dataArray2 store previous values of the values being plotted. These are used to erase the previous points from the screen when the graph cycles back to the right side of the screen. They are dimensioned form 0 to the maximum x coordinate of the graph, which in this example is the screen width. The variables x, y, and r are temporary values used for counting and storing values during computations.

The plotting is done within the loop routine. The first two lines are commands to put a black dot in the previous position plotted. This is to erase the previous line when the graph has reached the end of the graph and cycled back to the beginning. There are two of them because we are plotting two variables in this example.

Next the analog port is read and the value stored in the variable r by the line r=analogRead(A4). The map function (built into the C language used by this compiler) is then used to convert this value to a screen coordinate, which is stored in the variable y. This value is stored in the dataArray array by the command dataArray[dataPoint]=y. The line Tft.fillCircle(dataPoint, y, dotSize, CYAN) then puts a dot of size dotSize at the x and y coordinates dataPoint, y using color CYAN. Of course, you could use any color you want. Since this example plots two values, the process is repeated for the second reading. The only change in the second set of instructions is to replace A4 with A5, dataArray with dataArray2, and the color to be used to plot with from CYAN to BRIGHT_RED. Of course, again, you cold use any color you want.

After the two points have been plotted, the variable dataPoint is increased by 1 using the command dataPoint++, so that the next point will be plotted one point to the right of the current point. For those of you not familiar with dataPoint++, this is equivalent to the command dataPoint=dataPoint+1 used in some older languages like BASIC.

If the value of dataPoint has increased enough to reach or exceed the limit screenWidth, the if statement resets it to 0. At that point the graph would start at the left side of the screen again. As explained, the fact that the line is starting again at the left of the screen is why the previous points are erased by the two fillCircle commands at the beginning of the loop plotting black dots in the previous positions. Figure 10.1 is a photo of the screen.

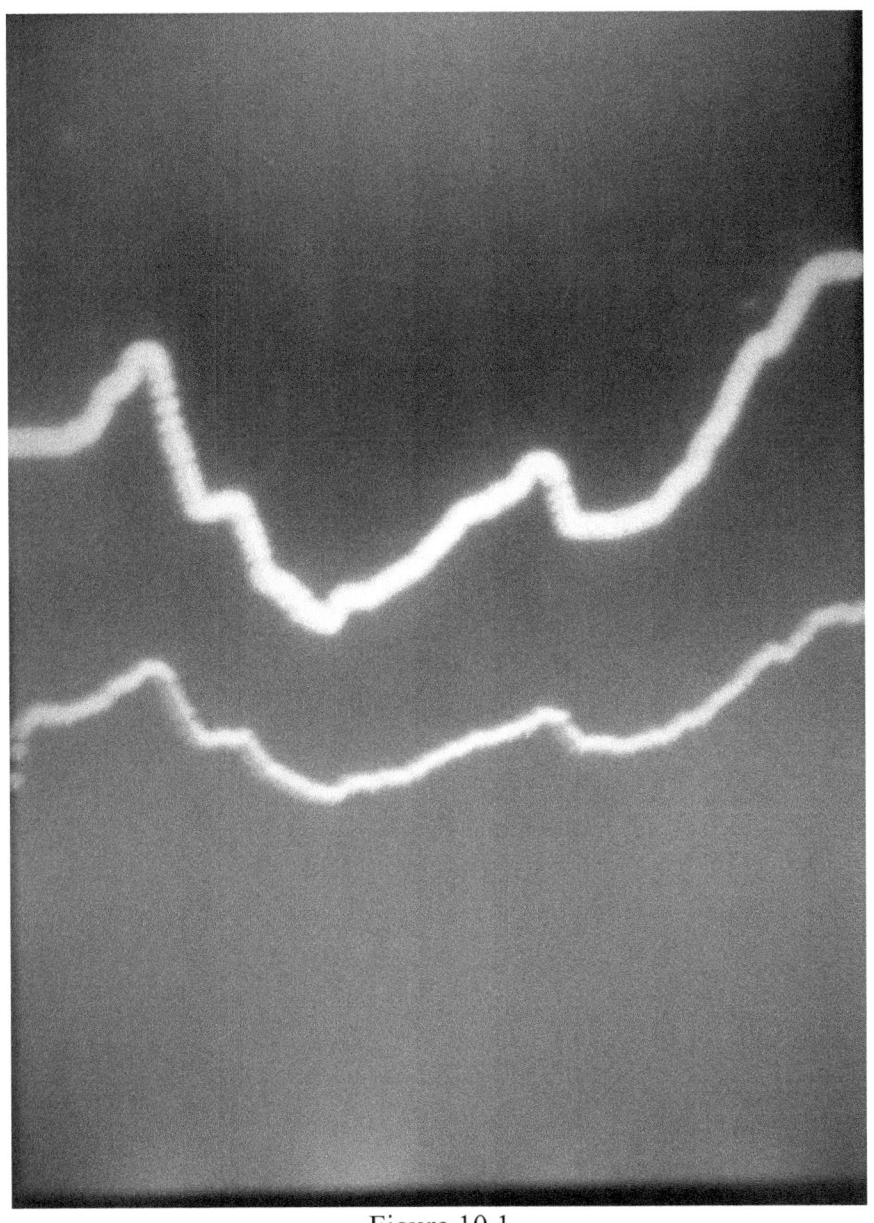

Figure 10.1

Sketch 10.1 drew the graph in portrait mode. That is, with the graph progressing across the width of the screen. That has the advantage that you can have buttons or text below the graph. If you want the graph go be displayed in landscape mode, with the X axis going across the width of the screen, Sketch 10.2 shows the necessary modifications.

```
#include <SeeedTouchScreen.h>
#include <TFTv2.h>
#include <SPI.h>

const int screenHeight = 320;
const int screenWidth = 240;
const int MinimumPressure = 150;
const int PlotTop = screenWidth-10;
const int PlotBottom = 10;
const int plotEnd = screenHeight - 10;
const long intervalTime = 100;
const int dotSize = 1;
int minValue = 0;
int maxValue = 1023;
int dataPoint=0;
int dataArray[plotEnd];
int dataArray2[plotEnd];
int x;
int y;
int r;
TouchScreen ts = TouchScreen(XP, YP, XM, YM);

void setup() {
   Tft.TFTinit();
   TFT_BL_ON;
}

void loop() {
  Tft.fillCircle(dataArray[dataPoint],dataPoint,dotSize,BLACK);
  Tft.fillCircle(dataArray2[dataPoint],dataPoint,dotSize,BLACK);
  r=analogRead(A4);

  y=map(r,maxValue,minValue,PlotTop,PlotBottom);
  dataArray[dataPoint]=y;
  Tft.fillCircle(y,dataPoint,dotSize,CYAN);

  r=analogRead(A5);
  y=map(r,maxValue,minValue,PlotTop,PlotBottom);
  dataArray2[dataPoint]=y;
```

```
Tft.fillCircle(y,dataPoint,dotSize,BRIGHT_RED);

  dataPoint++;
  if (dataPoint>=plotEnd){
    dataPoint=0;
  }
  delay(intervalTime);
}
```

Sketch 10.2

Most of the modifications consist of switching variables around, like swapping the order of dataPoint and y or minValue and maxValue so that the x and y axis is swapped. The basic explanation given for Sketch 10.1 therefore applies to Sketch 10.2. One change that is a real difference is that a new variable, plotEnd, has been added. In Sketch 10.1, the graph went all the way from the left side of the screen to the right side, and you had an option to put buttons or text below the graph. In Sketch 10.2, the graph goes from what would normally be the top of the screen to the bottom if you were holding the screen vertically. Putting plotEnd in the code allows you to end the plot before the bottom of the screen, so you can still put something there. However, you should note that any text you put at the bottom of the screen would be perpendicular to the direction of the graph, since the drawString command prints text in the portrait mode. Figure 10.2 shows this.

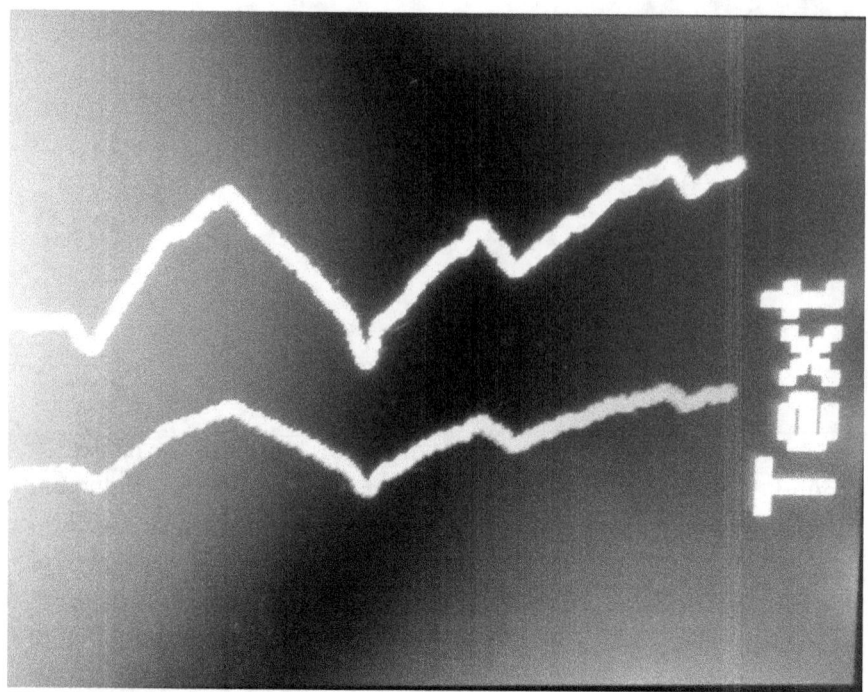

Figure 10.2

Note that the command to print the text is not included in Sketch 10.2, but it simply consists of adding
Tft.drawString("Text", 60, screenHeight - 40, 4, WHITE);
to the end of the setup routine.

The problem mentioned above with the text direction refers to the version 2 of the Seeed screen. The Seeed touch screen model 1 actually has a command in its library that allows you rotate text. This command is Tft.setDisplayDirect(direction), where you replace "direction" with LEFT2RIGHT, RIGHT2LEFT, DOWN2UP, or UP2DOWN. Sketch 10.3 shows the code for the landscape mode plotter with text using the version 1 touch screen.

```
#include <TouchScreen.h>
#include <TFT.h>

const int screenHeight = 320;
const int screenWidth = 240;
const int MinimumPressure = 150;
const int plotTop = screenWidth-10;
const int plotBottom = 50;
```

```
const int plotEnd = screenHeight;
const long intervalTime = 100;
const int dotSize = 1;
int minValue = 0;
int maxValue = 1023;
int dataPoint=0;
int dataArray[plotEnd];
int dataArray2[plotEnd];
int x;
int y;
int r;

void setup()
{
  Serial.begin(9600);
  Tft.init();  //init TFT library
  Tft.setDisplayDirect(UP2DOWN);
  Tft.drawString("Text here",30,20,4,WHITE);
}

void loop(){
  Tft.fillCircle(dataArray[dataPoint],dataPoint,dotSize,BLACK);
  Tft.fillCircle(dataArray2[dataPoint],dataPoint,dotSize,BLACK);
  r=analogRead(A4);
  y=map(r,maxValue,minValue,plotTop,plotBottom);
  dataArray[dataPoint]=y;
  Tft.fillCircle(y,dataPoint,dotSize,CYAN);

  r=analogRead(A5);

  y=map(r,maxValue,minValue,plotTop,plotBottom);
  dataArray2[dataPoint]=y;
  Tft.fillCircle(y,dataPoint,dotSize,WHITE);

  dataPoint++;
  if (dataPoint>=plotEnd){
    dataPoint=0;
  }
  delay(intervalTime);
}
```

Sketch 10.3

This plots the two graphs in landscape more with "Text here" beneath the graph. The command
Tft.setDisplayDirect(UP2DOWN);
Points the text in the right direction, and
Tft.drawString("Text here",30,20,4,WHITE);
prints the text at vertical position 30, horizontal position 20, size 4, color WHITE. You can fiddle with these values as you like. In particular, you will want to vary the horizontal position to center the text depending on how long it is.

Chapter 11

Changing Screens

In previous chapters, we have discussed many different types of screen displays. It would be nice if you could have your sketch display several different screens with different functions and switch between them. In this chapter, we will discuss doing that.

To have multiple possible screens, you need to number them and have a mechanism to switch screens. The simplest mechanism is to have buttons on the screen to change the screens, preferably to go up to a higher screen number and one to go down to a lower one. For this example, we will assume that there are three screens, one with sliders as described in the last chapter, one with a numeric keypad as discussed in chapter 8, and one with a scrollable menu as discussed in chapter 6. The following code is one way to do this. First, we need to define the variables used to control the screens, as follows.

```
int numScreens = 3;
int screenButtonColor = CYAN;
int screenButtonTextColor = BLACK;
int screenButtonTextSize = 3;
int screenButtonTextHeight = 7.5 * screenButtonTextSize;
int screenButtonTextWidth = 6 * screenButtonTextSize;
int screenButtonWidth = 30;
int screenButtonHeight = 50;
int screenButtonTop[] = {100,170};
int screenButtonLeft[] = {210,210};
char* screenButtonText[] = {">","<"};
int whichScreen = 1;
```

The variable numScreens is the number of screens your sketch has. The other variables are mostly self explanatory, giving the colors of the buttons and text within them, the size of the text, and the size and position of the screens. Most of these will be the same for the two buttons, so they do not need arrays. The top and left position of the buttons will be different (at least one of these qualities), and may even change within the program (as will be explained later. The "text" of the buttons will actually be arrows, and we will use the < and > symbols for this. The variable whichScreen is the current screen being displayed, and will be from 1 to numScreens.

We need a subroutine to display the buttons, which will be

```
void displayScreenButtons() {
  int x;
  for (x=0;x<2;x++) {
    Tft.fillRectangle(screenButtonLeft[x], screenButtonTop[x], screenButtonWidth, screenButtonHeight,screenButtonColor);
    Tft.drawString(screenButtonText[x], screenButtonLeft[x] + (screenButtonWidth-(screenButtonTextWidth))/2, screenButtonTop[x] + (screenButtonHeight-screenButtonTextHeight)/2 , screenButtonTextSize,screenButtonTextColor);
  }
}
```

The for loop simply counts through the two buttons. The fillRectangle command draws a box, and the drawString command puts the < or > symbol in the box. It is worth noting that several functions in this book, such as scrolling text, clear the screen. You will want to redisplay the buttons every time you do this to make sure these buttons are always on the screen, so it is useful to have a subroutine that clears the screen and automatically restores the screen change buttons, like this.

```
void clearScreen(){
  Tft.fillRectangle(0, 0, screenWidth, screenHeight, BLACK);
  displayScreenButtons();
}
```

This simply fills the screen with a large black rectangle and then displays the which screen buttons. Next you need a subroutine to display the current screen based on which screen is currently selected. This is quite simple.

```
void displayCurrentScreen() {
  clearScreen();
```

```
if (whichScreen == 1) {
 displaySliders();
  }
if (whichScreen == 2) {
 displayTextMenu();
  }
if (whichScreen == 3) {
 displayNumButtons();
  }
 displayScreenButtons();
}
```

This clears the screen, then displays one of the possible screens based on what the whichScreen variable is. After displaying the screen, it redisplays the buttons. Of course, the buttons were displayed by the clearScreen subroutine, but calling for the buttons to be displayed again makes sure that they were not overwritten when the screen is displayed. Of course, this is just an example, the actual screens displayed might not be the slider, text menu, or number buttons screens.

Now let's get into how you change the screens. This code goes within the loop routine.

```
Point p = ts.getPoint();
//map the ADC value read to into pixel co-ordinates
p.x = map(p.x, TS_MINX, TS_MAXX, 0, 240);
p.y = map(p.y, TS_MINY, TS_MAXY, 0, 320);
//  Which Screen stuff here

if (p.z > MinimumPressure && p.z<10000) {
    px = p.x;
    py = p.y;

      if (py > screenButtonTop[0] && py < screenButtonTop[0] + screenButtonHeight) {
        if (px > screenButtonLeft[0] && px < screenButtonLeft[0] + screenButtonWidth) {
           whichScreen = whichScreen + 1;
           if (whichScreen > numScreens) {
```

```
      whichScreen = 1;
    }
    displayCurrentScreen();
    delay(750);
    }
  }

    if (py > screenButtonTop[1] && py < screenButtonTop[1] +
screenButtonHeight) {
    if (px > screenButtonLeft[1] && px < screenButtonLeft[1] +
screenButtonWidth) {
      whichScreen = whichScreen - 1;
      if (whichScreen < 1) {
        whichScreen = numScreens;
      }
      displayCurrentScreen();
      delay(750);
    }
    }
} // End screen selection here
```

The first if statement, of course, checks to see if the screen is being pressed at all. If so, it records the x and y positions of the touch in px and py. The next if statement checks to see if the touch is within the range from the top to the bottom of the first button (button 0). The next if statement sees if the touch is within the range from the left to the right of that button. These two statements confirm that the touch is actually on the first button. All this is the same as any button, as described in Chapter 8. If the touch is on the first button, whichScreen increases by 1. If the new value is greater than the number of screens, the if (whichScreen > numScreens) statement immediately after whichScreen = whichScreen + 1 resets it to 1. Thus, the number of the screen wraps around. Once the screen number is determined, the next statement calls the displayCurrentScreen subroutine to display the newly selected screen. Then there is a delay. This allow you to take your finger off the screen before the sketch loops around again. If you did not have this delay, the screens would rapidly shift each time the code got around to this statement, because your finger is still on the button. This is actually good, because it means that if you hold down the button, the screen will continue to change. Then you can release the button when the screen you want is shown. Press the button briefly to move to the next screen, hold it down to keep changing the screen.

After this code that increases the screen number if you press the top button (marked >), you have basically the same code for the bottom button. This code checks to see if the touch is within the range of the top and bottom of the button and the left and right of the button. If so, it decreases the screen number by 1. If the new number is less than 1 (the lowest screen number), it resets the value to numScreens (the highest screen number). Again, it then displays the new screen and delays to give you time to take your finger off the screen.

Now you have to have the sketch respond differently for each screen. You do this simply by putting the code for each function, such as responding to a touch on a slider or a text menu, inside a conditional statement such as
if (whichScreen == 1) {

Sketch 11.1 gives an example of an entire operation.

```
#include <TFTv2.h>
#include <SPI.h>
#include <SeeedTouchScreen.h>

//General variables
int screenHeight = 320;
int screenWidth = 240;
int MinimumPressure = 50; //Minimum pressure to consider touch
int px;
int py;

//Which screen variables
int numScreens = 3;
int screenButtonColor = CYAN;
int screenButtonTextColor = BLACK;
int screenButtonTextSize = 3;
int screenButtonTextHeight = 7.5 * screenButtonTextSize;
int screenButtonTextWidth = 6 * screenButtonTextSize;
int screenButtonWidth = 30;
int screenButtonHeight = 50;
```

```
int screenButtonTop[] = {100,170};
int screenButtonLeft[] = {210,210};
char* screenButtonText[] = {">","<"};
int whichScreen = 1;

//Slider variables
float fractionValue;
int sliderFillColor = RED;
int sliderColor = CYAN;
int sliderTextSize = 2;
int sliderWidth[] = {30,30,30,30};
float sliderHeight[] = {260,260,260,260};
float sliderTop[] = {20, 20,20,20};
int sliderLeft[] = {5,55,105,155};
float sliderValue[] = {110,25,50,75};
float sliderMinValue[] = {100,0,0,0};
float sliderMaxValue[] = {200,100,100,100};
char* sliderTextLabel[] = {"A","B","C","D"};
int numSliders = 4;
int sliderSelection = -1;
int previousSliderSelection = -1;
int sliderStartTouch = -1;
int sliderTextColor = CYAN;

//Text menu items
char* textItem[] = {"Item 1","Item 2","Item 3","Item 4","Item 5","Item 6","Item 7","Item 8","Item 9","Item 10","Item 11","Item 12","Item 13","Item 14","Item 15"};
int numTextMenuItems = 15;
int menuTextSize = 4;
int menuTextSpacing = 5;
int menuTotalTextHeight = 7.5 * menuTextSize + menuTextSpacing;
int menuMaxItems = screenHeight/(menuTotalTextHeight);
int itemNumber;
int startItem = 0; // Top item shown
int previousStartItem = 0;
int UpDown; // Scroll up or down
int menuRightMargin = screenWidth * .85;
int lineTouched; // Line on screen touched
```

```
int textMenuSelection = -1; //Item picked from list, initial to less
than any item
boolean textMenuSelectionMade = false; //Was an item picked?
int selectedColor = WHITE;
int menuTextColor = CYAN;
int printColor;

// Number keypad variables
int numButtonWidth = 50;
int numButtonHeight = 40;
int numButtonHSpace = 10;
int numButtonVSpace = 15;
int topSpace = 40;
int numberButtonsTextSize = 4;
int numbersTextSize = 4;
int numbersTextHeight = 7.5 * numbersTextSize;
char* numButtonText[] =
{"1","2","3","4","5","6","7","8","9","E","0","C"};
String combined = "";
String oneChar;
int numberStartTouch = -1;
int numberSelection = -1;
int lenCombined = 0;
long numberValue;

TouchScreen ts = TouchScreen(XP, YP, XM, YM); //init
TouchScreen port pins

void setup() {
 Serial.begin(9600);
 TFT_BL_ON;
 Tft.TFTinit();
 Tft.fillRectangle(0, 0, screenWidth, screenHeight, BLACK);
 whichScreen = 1;
 displayCurrentScreen();
}

void loop() {
  int x;
  int y;
```

```
Point p = ts.getPoint();
//map the ADC value read to into pixel co-ordinates
p.x = map(p.x, TS_MINX, TS_MAXX, 0, 240);
p.y = map(p.y, TS_MINY, TS_MAXY, 0, 320);
//  Which Screen stuff here

if (p.z > MinimumPressure && p.z<10000) {
    px = p.x;
    py = p.y;

    if (py > screenButtonTop[0] && py < screenButtonTop[0] +
screenButtonHeight) {
        if (px > screenButtonLeft[0] && px < screenButtonLeft[0] +
screenButtonWidth) {
            whichScreen = whichScreen + 1;
            if (whichScreen > numScreens) {
              whichScreen = 1;
            }
            displayCurrentScreen();
            delay(750);
        }
    }

    if (py > screenButtonTop[1] && py < screenButtonTop[1] +
screenButtonHeight) {
        if (px > screenButtonLeft[1] && px < screenButtonLeft[1] +
screenButtonWidth) {
            whichScreen = whichScreen - 1;
            if (whichScreen < 1) {
              whichScreen = numScreens;
            }
            displayCurrentScreen();
            delay(750);
        }
    }

} // End screen selection here

// Slider handling stuff here
```

```
if (whichScreen == 1) {
 if (p.z == 0 && sliderStartTouch == 0) {
  if (sliderSelection > -1){
    if (sliderSelection  == 0) {
     Serial.print ("First =");
     Serial.println (sliderValue[sliderSelection]);
    }
    if (sliderSelection == 1) {
     Serial.print ("Second =");
     Serial.println (sliderValue[sliderSelection]);
    }
    if (sliderSelection == 2) {
     Serial.print ("Third =");
     Serial.println (sliderValue[sliderSelection]);
    }
    if (sliderSelection == 3) {
     Serial.print ("Fourth = ");
     Serial.println (sliderValue[sliderSelection]);
    }
    if (sliderSelection == 4) {
     Serial.print ("Fifth = ");
     Serial.println (sliderValue[sliderSelection]);
    }
    sliderSelection = -1;
    previousSliderSelection = -1;
  }
  sliderStartTouch = -1;
  delay (100);
   }
  if (p.z > MinimumPressure && p.z<10000) {
    if (sliderStartTouch == -1) {
     sliderStartTouch = 0;
     delay (100);
    }
    px = p.x;
    py = p.y;
    sliderSelection = -1;
    for (x=0;x<numSliders;x++) {
       if (py > sliderTop[x]-20 && py < sliderTop[x] + sliderHeight[x]+20) {
```

```
            if (px > sliderLeft[x] && px < sliderLeft[x] +
sliderWidth[x]) {
              sliderSelection = x;
              sliderValue[x] = int(((sliderTop[x] + sliderHeight[x]) -
py) /sliderHeight[x] * (sliderMaxValue[x]-sliderMinValue[x]) +
sliderMinValue[x]);
              if (sliderValue[x] < sliderMinValue[x]) {sliderValue[x]
= sliderMinValue[x];}
              if (sliderValue[x] > sliderMaxValue[x]) {sliderValue[x]
= sliderMaxValue[x];}
           }
          }
        }
      //if (previousSliderSelection != -1){
      if (sliderSelection != -1){
        Serial.println("Slider");
        displaySliders();
      }
      previousSliderSelection = sliderSelection;
    }
  } // End Slider stuff
  //Number keypad stuff here
  if (whichScreen == 3) {

    if (p.z == 0 && numberStartTouch == 0) {
      if (numberSelection > -1){
        if (numberSelection < 9 || numberSelection == 10) {
        combined = combined + numButtonText[numberSelection];
        lenCombined = lenCombined + 1;
        char toDisplay[lenCombined+1];
        combined.toCharArray(toDisplay,lenCombined+1);
        Tft.drawString(toDisplay, 0, 0, numbersTextSize,WHITE);
        }
        if (numberSelection == 11) {
        lenCombined = lenCombined - 1;
        combined = combined.substring(0,lenCombined);
        // Erase previous string
        Tft.fillRectangle(0, 0, screenWidth, numbersTextHeight,
BLACK);
        char toDisplay[lenCombined + 1];
```

```
      combined.toCharArray(toDisplay,lenCombined + 1);
      Tft.drawString(toDisplay,0, 0, numbersTextSize,WHITE);
     }
    if (numberSelection == 9) {
      char toDisplay[lenCombined + 1];
      combined.toCharArray(toDisplay,lenCombined + 1);
      numberValue = atol(toDisplay);
       Serial.print ("combined = ");
       Serial.println (numberValue);
       combined="";
       lenCombined = 0;
       Tft.fillRectangle(0, 0, screenWidth, numbersTextSize, BLACK);
     }
      numberSelection = -1;
    }
    numberStartTouch = -1;
    delay (100);
   }
  if (p.z > MinimumPressure && p.z<10000) {
     if (numberStartTouch == -1) {
       numberStartTouch = 0;
       delay (100); // Prevent bounce. may vary time
     }
     px = p.x;
     py = p.y;
     for (x=0;x<3;x++) {
       for (y=0;y<4;y++) {
         if (px > x * numButtonWidth + (x + 1) * numButtonHSpace && px < x * numButtonWidth + (x + 1) * numButtonHSpace + numButtonWidth){
           if (py >  y * numButtonHeight + (y + 1) * numButtonVSpace + topSpace && py <  y * numButtonHeight + (y + 1) * numButtonVSpace + topSpace + numButtonHeight) {
             numberSelection = 3 * y + x;
           }
         }
       }
     }
     delay (100); // Prevent bounce. may vary time
```

```
    }
  }// End number keypad here

  // Text menu
  if (whichScreen == 2) {
  if (p.z > MinimumPressure && p.z<10000) {
     if(p.y < menuTotalTextHeight && p.x > menuRightMargin) {UpDown = -1;}
     if (p.y > screenHeight - menuTotalTextHeight && p.x > menuRightMargin) {UpDown=1;}
     if (p.x<menuRightMargin) {
       lineTouched = p.y / menuTotalTextHeight;
       textMenuSelection = lineTouched + startItem;
       textMenuSelectionMade = true;
       displayTextMenu();
     }
     delay (500); // Prevent bounce. may vary time
  }
  //Try to prevent continuous input
  if (textMenuSelectionMade && p.z == 0) {
    Serial.println (textItem[textMenuSelection]);
    textMenuSelectionMade = false;
  }
  if (UpDown!=0) {
    if (p.z == 0) {
      startItem=startItem-UpDown;
      //Do not let start item be less than first item
      if (startItem<0) {
        startItem = 0;
      }
      if (startItem>numTextMenuItems - menuMaxItems) {
        startItem = numTextMenuItems - menuMaxItems;
      }
      UpDown=0;
      displayTextMenu();
    }
  }
  } // End Text menu
}
```

```
void clearScreen(){
 Tft.fillRectangle(0, 0, screenWidth, screenHeight, BLACK);
 displayScreenButtons();
}

void displayCurrentScreen() {
 clearScreen();
 if (whichScreen == 1) {
  displaySliders();
   }
 if (whichScreen == 2) {
   displayTextMenu();
   }
 if (whichScreen == 3) {
   displayNumButtons();
   }
 displayScreenButtons();
}

void displayScreenButtons() {
 int x;
 for (x=0;x<2;x++) {
    Tft.fillRectangle(screenButtonLeft[x], screenButtonTop[x], screenButtonWidth, screenButtonHeight,screenButtonColor);
    Tft.drawString(screenButtonText[x], screenButtonLeft[x] + (screenButtonWidth - (screenButtonTextWidth))/2, screenButtonTop[x] + (screenButtonHeight - screenButtonTextHeight)/2 , screenButtonTextSize, screenButtonTextColor);
   }
}

void displaySliders() {
  int x;
  for (x=0;x<numSliders;x++) {
     fractionValue = (sliderValue[x] - sliderMinValue[x])/ (sliderMaxValue[x] - sliderMinValue[x]);
     Tft.fillRectangle(sliderLeft[x], sliderTop[x], sliderWidth[x], sliderHeight[x] * (1 - fractionValue), sliderColor);
      if (fractionValue > 0){
```

```
      Tft.fillRectangle(sliderLeft[x], sliderTop[x] + (1-fractionValue)
* sliderHeight[x], sliderWidth[x], sliderHeight[x] * fractionValue,
sliderFillColor);
    }
    Tft.drawString(sliderTextLabel[x], sliderLeft[x], sliderTop[x] +
sliderHeight[x] + 3, sliderTextSize, sliderTextColor);
  }
}

void displayTextMenu() {
  int x;
  if (startItem != previousStartItem) {
    clearScreen();
    previousStartItem = startItem;
  }
  Tft.fillRectangle(menuRightMargin, 0, screenWidth,
menuTotalTextHeight, BLUE);
  Tft.fillRectangle(menuRightMargin, screenHeight -
menuTotalTextHeight, screenWidth, screenHeight, BLUE);
  for (x=0;x<menuMaxItems;x++) {
    if (x + startItem == textMenuSelection) {
      printColor = selectedColor;
    }
    else {
      printColor = menuTextColor;
    }
    Tft.drawString(textItem[x + startItem], 0, x *
menuTotalTextHeight, menuTextSize, printColor);
  }
}

void displayNumButtons() {
  int x;
  int y;
  for (x=0;x<3;x++) {
    for (y=0;y<4;y++) {
      Tft.fillRectangle(x * numButtonWidth + (x + 1) *
numButtonHSpace, y * numButtonHeight + (y + 1) *
numButtonVSpace + topSpace, numButtonWidth,
numButtonHeight,CYAN);
```

```
    Tft.drawString(numButtonText[3 * y + x], x * numButtonWidth
+ (x + 1) * numButtonHSpace + numButtonWidth/3, y *
numButtonHeight + (y + 1) * numButtonVSpace + topSpace +
numButtonHeight * .15, numberButtonsTextSize, BLACK);
   }
 }
 char toDisplay[lenCombined+1];
 combined.toCharArray(toDisplay,lenCombined+1);
 Tft.drawString(toDisplay, 0, 0, numbersTextSize, WHITE);
}
```

Sketch 11.1

Here we have added the routines for sliders, a scrollable text menu, and a numeric keypad. The basic routines are the same as in the chapters that discussed these functions. I have made a few changes. The main change is that I have changes the variable names to make then unique to each function.. For example, for the sliders variable, I have added "slider" to the beginning of variable names, such as sliderFillColor and sliderTextSize, rather than fillColor and textSize. Also, many variables like x are defined within a subroutine rather than global. These steps insures that the code does not accidentally change a variable used by one function while running another, and allows each to have its own characteristics like text size and color.

I have also added some extra code to some of the subroutines that display the screens. For example, for the numeric keypad, I added code to draw the current number value at the top to the displayNumButtons subroutine. This was not necessary in the original code, because displayButtons was called only once, in the setup routine. The numbers were then manipulated within the loop and were displayed then. However, in this sketch, the numeric keypad will be erased when you switch to another screen and then redrawn using the displayButtons subroutine when you switch back to it. If the displayButtons subroutine does not restore the numbers to the screen, you will not see them when you switch back to this screen. This is just an example of the extra thought you may need to put into this idea of having the display switch among several screens. However, it is worth the effort to maximize the usefulness of the touch screen with the Arduino.

Chapter 12

Using the Micro SD Card

Many touch screens come with an SD or micro SD card slot that the touch screen can access. The main reason for this is that one common use of the graphic screen is to display pictures, and a detachable memory like an SD card is the easiest way to load a bunch of pictures to view on the screen. The Arduino has limited memory, and could only hold a limited number of pictures. While holding pictures for display is nice, you can also use that SD card to hold data from your project and easily transfer it. For example, suppose you are using the Arduino to collect data like weather conditions or to monitor a scientific experiment. Storing the data on the card, which you can easily remove from the touch screen and transfer to your computer, can be useful, especially if the Arduino is distant from a Wi-Fi connection or other means of transferring the data. In this chapter, I will discuss using the SD card for graphics and also for storing data.

As I noted in Chapter 2, you will need to download the file SD-master.zip and install it on your computer. This file contains a library of examples for using the SD card. However, these examples are not written specifically for touch screens, and will not work with a touch screen without modification.

First, let's look at setting up the sketch, such as includes, variables and constants, and initialization.

```
#include <SPI.h>
#include <TFTv2.h>
#include <SeeedTouchScreen.h>
#include <SD.h>

const int PIN_SD_CS = 4; // pin of sd card
File dataFile;
int Message;
String fileName;

void setup(){
```

```
    pinMode(PIN_SD_CS,OUTPUT);
    digitalWrite(PIN_SD_CS,HIGH);
    Tft.TFTinit();
    Sd2Card card;
    card.init(SPI_FULL_SPEED, PIN_SD_CS);
    Tft.drawString("Initializing", 0, 0, 3, WHITE);
    if (!SD.begin(PIN_SD_CS)) {
      Tft.drawString("Failed", 0, 30, 3, WHITE);
      return;
    }
    Tft.drawString("Success", 0, 30, 3, WHITE);
}
```

First, you will see that SD.h has been added to the includes. This is necessary to use the SD card. The SeeedTouchScreen.h is not actually needed to use the SD card, but is needed for screen writing and touch detection.

PIN_SD_CS identifies the CS pin used by the SD card. This may vary with different touch screens, so we will use a named constant to make it easier to change. Also, using a name makes the code easier to understand. The line File dataFile defines dataFile as a file variable. The variable filename is the name of the file you will be saving the data in. The variable Message, which can be any variable type including string, is what is saved to the file.

In the setup routine, the lines
pinMode(PIN_SD_CS,OUTPUT);
 digitalWrite(PIN_SD_CS,HIGH);
initialize the CS pin. The lines
TFT_BL_ON;
Tft.TFTinit();
Turn on the backlight and initialize the touch screen, as they did in the previous programs. The lines
Sd2Card card;
card.init(SPI_FULL_SPEED, PIN_SD_CS);
define card as a card variable and initialize the card.

The statement SD.begin(PIN_SD_CS) activates the card communication, just as Serial.begin activates serial communication. SD.begin returns true if the communication has been successfully established. The Tft.drawString statements just report on the success of this. First, it displays "Initializing". The exclamation point means NOT, so !SD.begin(PIN_SD_CS) means SD.begin did not return true and the communication was not activated. In that case, the sketch sends the "Failed" message to the screen and shuts down the process with the return command. Otherwise, the sends the "Success" message to the screen. This is not necessary, of course, but it does help with debugging and confirms that your Arduino is working properly when you start it.

Now that we have the SD card initialized, let's look at saving data to the card. In order to keep it organized and flexible, we will put the code for writing to the card in a subroutine.

```
void writeToFile(){
  dataFile = SD.open(fileName, FILE_WRITE);
  if (dataFile) {
   dataFile.println(Message);
   dataFile.close();
  }
  else {
   Tft.drawString("Write error", 0, 60, 3, RED);
  }
}
```

The first statement opens a file with the name given in the string variable fileName. This variable is assigned a string, such as "data.dat," before the writeToFile subroutine is called. It can be given one name that is never changed when it is declared at the beginning of the sketch, or it can be changed within the sketch, allowing the writeToFile subroutine to write any number of different files. The constant FILE_WRITE is a reserved keyword assigned within the SD.h file included at the beginning of the sketch, and tells the SD.open command to open the file for writing. The SD.open command returns a file identifier which is then stored in the File type variable dataFile. The if (dataFile) statement confirms that dataFile has been set. If so, the command dataFile.println(Message) prints the variable Message to the file (very similar to Serial.println) and then closes the file again. The variable Message can be a number or a string. A value must be given this variable before the writeToFile subroutine is called. If dataFile was no properly set, the else clause displays an error message on the screen.

There is one interesting thing about using the SD.open command. In many programming languages, opening a file using the WRITE mode automatically erases the existing file. In this language, the SD.open command does not erase the existing file, but appends to it. That is, the SD.open(filename, FILE_WRITE) command acts like opening a file in APPEND mode does in some other languages. If you want the process of writing to a file to start with a clean file even if the file already exists, you would need to put
SD.remove(filename);
before the SD.open command. The SD.remove command will delete the existing file so the new data being saved will be the entire file. In most cases, however, you would not want to do this, but would want the new data added to the existing file.

Let's take a look at an entire sketch using this setup. For this example, we will have the sketch save the value of two analog inputs in separate files once every five seconds.

```
#include <SPI.h>
#include <TFTv2.h>
#include <SeeedTouchScreen.h>
#include <SD.h>

const int PIN_SD_CS = 4; // pin of SD card
File dataFile;
int Message;
String fileName;

void setup(){
  pinMode(PIN_SD_CS,OUTPUT);
  digitalWrite(PIN_SD_CS,HIGH);
  Tft.TFTinit();
  Sd2Card card;
  card.init(SPI_FULL_SPEED, PIN_SD_CS);
  Tft.drawString("Initializing", 0, 0, 3, WHITE);
 if (!SD.begin(PIN_SD_CS)) {
  Tft.drawString("Failed", 0, 30, 3, WHITE);
  return;
 }
 Tft.drawString("Success", 0, 250, 3, WHITE);
}
```

```
void loop(){
  fileName = "A4.dat";
  Message = analogRead(A4);
  writeToFile();
  fileName = "A5.dat";
  Message = analogRead(A5);
  writeToFile();
  delay(5000);
}

void writeToFile(){
  dataFile = SD.open(fileName, FILE_WRITE);
  if (dataFile) {
    dataFile.println(Message);
    dataFile.close();
  }
  else {
    Tft.drawString("Write error", 0, 60, 3, RED);
  }
}
```

Sketch 12.1

    Here we have the setup and inititialization as described previously, plus something in the loop routine. First, it sets the file name to "A4.dat." Then it sets the variable Message to the value of the current analog input A4. Then it calls the writeToFile subroutine. It then repeats the process for the A5 analog input, using a different file name. It then delays 5000 milliseconds (5 seconds) before continuing in the loop.

    This is a simple example of saving data to the SD card, intended for demonstration purposes. It does have a serious flaw. You have no control over it. As long as the Arduino has power, it keeps writing the data to the card. The only way to stop it is to unplug the Arduino from the power supply. If it happened to be writing at that moment, it might even corrupt the file. We need more control, like some buttons to turn the writing of the data on and off. Sketch 12.2 adds in some buttons to do this.

```cpp
#include <SPI.h>
#include <TFTv2.h>
#include <SeeedTouchScreen.h>
#include <SD.h>

// Buttons variables
int screenHeight = 320;
int screenWidth = 240;
int MinimumPressure = 50;
int textSize = 3;
int textHeight = 7.5 * textSize;
int textWidth = 6 * textSize;
int stringLength;
String S;
int x;
int y;
int px;
int py;
int selection = -1;
int previousSelection = -1;
int startTouch = -1;
int buttonColor;
int textColor = BLACK;
int normalButtonColor=CYAN;
int activeButtonColor=RED;
int buttonWidth[] = {200,200};
int buttonHeight[] = {40,40};
int buttonTop[] = {70, 140};
int buttonLeft[] = {10,10};
char* buttonText[] = {"Start", "Stop"};
int numButtons = 2;
bool saving = false;

// Timer
unsigned long previousMillis = 0;
const long interval = 5000; // interval between writes (milliseconds)

TouchScreen ts = TouchScreen(XP, YP, XM, YM);
```

```
const int PIN_SD_CS = 4;
File dataFile;
int Message;
String fileName;

void setup(){
  pinMode(PIN_SD_CS,OUTPUT);
  digitalWrite(PIN_SD_CS,HIGH);
  Tft.TFTinit();
  Sd2Card card;
  card.init(SPI_FULL_SPEED, PIN_SD_CS);
  Tft.drawString("Initializing", 0, 0, 3, WHITE);
 if (!SD.begin(PIN_SD_CS)) {
  Tft.drawString("Failed", 0, 30, 3, WHITE);
  return;
 }
 Tft.drawString("Success", 0, 30, 3, WHITE);
 displayButtons();
}

void loop(){
  Point p = ts.getPoint();
  p.x = map(p.x, TS_MINX, TS_MAXX, 0, 240);
  p.y = map(p.y, TS_MINY, TS_MAXY, 0, 320);

  if (p.z == 0 && startTouch == 0) {
   if (selection > -1){
     if (selection == 0) {
      saving = true;
      Tft.drawString("Stopped",50,200,3,BLACK);
      Tft.drawString("Writing",50,200,3,BLUE);
     }
     if (selection == 1) {
      saving = false;
      Tft.drawString("Writing",50,200,3,BLACK);
      Tft.drawString("Stopped   ",50,200,3,BLUE);
     }
     selection = -1;
     previousSelection = -1;
     displayButtons();
```

```
      }
    startTouch = -1;
    delay (100);
    }
  if (p.z > MinimumPressure && p.z<10000) {
    if (startTouch == -1) {
      startTouch = 0;
      delay (100);
      }
    px = p.x;
    py = p.y;
    selection = -1;
    for (x=0; x<numButtons; x++) {
        if (py > buttonTop[x] && py < buttonTop[x] +
buttonHeight[x]) {
          if (px > buttonLeft[x] && px < buttonLeft[x] +
buttonWidth[x]) {
            selection =  x;
          }
        }
      }
      if (previousSelection != selection){
        displayButtons();
      }
      previousSelection = selection;
    delay (100);
    }
  if (saving) {
    unsigned long currentMillis = millis();
    if(currentMillis - previousMillis >= interval) {
    previousMillis = currentMillis;
    fileName = "A4.dat";
    Message = analogRead(A4);
    writeToFile();
    fileName = "A5.dat";
    Message = analogRead(A5);
    writeToFile();
   }
 }
}
```

```
void writeToFile(){
 dataFile = SD.open(fileName, FILE_WRITE);
 if (dataFile) {
  dataFile.println(Message);
  dataFile.close();
  Serial.print(fileName);
  Serial.print(" = ");
  Serial.println(Message);
 }
 else {
  Tft.drawString("Write error", 0, 60, 3, RED);
 }
}

void displayButtons() {
 for (x=0; x<numButtons; x++) {
   S = buttonText[x];
   stringLength = S.length();
   buttonColor = normalButtonColor;
   if (selection == x) {buttonColor = activeButtonColor;}
   Tft.fillRectangle(buttonLeft[x], buttonTop[x], buttonWidth[x], buttonHeight[x], buttonColor);
   Tft.drawString(buttonText[x],buttonLeft[x] + (buttonWidth[x] - (stringLength * textWidth))/2, buttonTop[x] + (buttonHeight[x]-textHeight)/2 , textSize, textColor);
 }
}
```

Sketch 12.2

This takes Sketch 12.1 and adds buttons as described in Chapter 8. We have added the usual buttons variables, and the displayButtons subroutine. As you can see from the variables, there are two buttons, one labeled "Start" and one labeled "Stop." A Boolean variable saving has been added that will tell whether data is being saved at any given time. It is initially set to false, so data is not written to the file when you first turn on the Arduino. The displayButtons subroutine is called in the setup routine to display the buttons when the Arduino is powered up.

Within the loop routine, the sketch still reads the analog inputs and saves the readings to the SD card. However, this code is within an if (saving) condition, so it only saves data if the saving Boolean variable is true. Another difference you should notice is that the delay command is missing. Instead, the sketch uses the millis function built into the Arduino language to check the time. It stores the current time in the variable currentMillis. Then it compares this time to the time that data was most recently saved, stored in the variable previousMillis. If the time difference is greater than the time period stored in the variable interval, the code changes previousMillis to equal currentMillis and writes the data to the SD card. This is necessary because the delay command stops the program flow entirely so no other commands are executed. If you used the delay command as we did in Sketch 12.1, you would have to hold down the Stop button for up to five seconds so that your finger would still be on that button when the program ended the delay and checked to see if it was being pressed. Using the timer function millis enables the rest of the code to execute while it is waiting to write to the SD card again, so your touch on the Stop button is detected immediately.

The code that detects the button touch is basically the same as in Chapter 8. The part that detects your selection when the button is actually being touched is the same. The difference is in how it responds to your selection when the button is release (when $p.z = 0$). If button 0 (the "Start" button) is pressed, it sets the variable saving to true, so that the data can be written to the SD card. If button 1 is pressed, it sets saving to false. These also send the word "Writing" or "Stopped" to the screen to keep you informed of the status of the saving. You probably noticed that before writing the word for the current status in blue, it writes the previous word in black. This erases the previous word from the screen. In previous chapters, we simply drew a black rectangle over the entire area where the previous text had been. This is necessary if you do not know what the previous text was, like in the numeric buttons chapter, or if there is a large area to be erased. However, if you know exactly what is written, as you do in this case, writing the old word in black (assuming that black is your background color) on the same spot has the effect of erasing it from the screen. It is nice to have options, so I thought I would throw a little twist in there for you.

Now that we have the ability to write to the card, it might

also be nice to be able to read. Of course, for most purposes where you have written to the card, the simplest thing to do might be to remove the card from the Arduino and put it in a card reader connected to a computer so you can use more complex software, like a spreadsheet, to read the data. However, you might want to read the data directly on the Arduino touch screen. Here is a subroutine that reads a text file.

```
void readFile(){
  int data;
  File dataFile;
  dataFile = SD.open(fileName);
  if (dataFile) {
    // read from the file until there's nothing else in it:
    while (dataFile.available()) {
      data = dataFile.read();
      // Do something with this data
      Serial.println(data);
    }
    dataFile.close();
  }
  else {
    // if the file didn't open, print an error:
    Tft.drawString("Cannot read", 10, 90, 3, RED);
  }
}
```

Here we initialize a variable data as an integer to receive the data and the File type variable dataFile to be the file. We use SD.open to open the file. The string variable fileName must have already been given a file name, of course. Note that this time, since we are reading instead of writing, the SD.open function does not include the FILE_WRITE constant. You could include FILE_READ in its place, but that is the default so it is not necessary.

The if (dataFile) test checks to see if the file was properly opened. If true, the code starts reading the data. If false, the "Cannot read" message is displayed on the screen.

The available() function is dataFile.available() is the same function often used as Serial.available(). It reports true if there is any data still to be read. If so, as long as there is data, the dataFile.read()

function reads the data. Here we are assuming that the data is a series of integer numbers. You can do whatever you want with the numbers, such as store them in an array. For this example, we are simply writing them out to the serial port.

The above example assumes that the information in the file is a series of integer numbers. If the data file contains a series of strings, a few modifications are necessary. These are shown in the following subroutine.

```
void readFile(){
  String data;
  File dataFile;
  dataFile = SD.open(fileName);
  Serial.println(fileName);
  if (dataFile) {
    // read from the file until there's nothing else in it:
    while (dataFile.available()) {
      data = dataFile.readStringUntil('\n');
      // Do something with this data
      Serial.println(data);
    }
    dataFile.close();
  }
  else {
    // if the file didn't open, print an error:
    Tft.drawString("Cannot read", 10, 90, 3, RED);
  }
}
```

There are really only two changes. First, the variable data is declared as a string. Second, dataFile.read() has been replaced by dataFile.readStringUntil('\n'). This reads data as a string until it encounters an end of line character in the file. The '\n' is a notation that means "end of line." Between the two codes, you can retrieve either numeric or sting data saved in a file on the SD card.

As an example, let's read in a text file from the card and use the scrolling text procedure from Chapter 5 to display it on the screen. Here is the code.

```
#include <SPI.h>
```

```
#include <TFTv2.h>
#include <SeeedTouchScreen.h>
#include <SD.h>

// card variables and constant
const int PIN_SD_CS = 4;
String fileName = "strings.txt";

//Scrolling text variables
int screenHeight = 320;
int screenWidth = 240;
int textSize = 4;
int textSpacing = 5;
int textColor = CYAN;
int totalTextHeight = 7.5 * textSize + textSpacing;
int maxItems = screenHeight/(totalTextHeight); // Maximum number of lines that will fit on screen
int numItems = 0;
int i;
int MinimumPressure = 50;
int startItem = 0;
int UpDown;
int rightMargin = screenWidth * .8;
String data[50];

TouchScreen ts = TouchScreen(XP, YP, XM, YM);

void setup(){
  Serial.begin(9600);
    pinMode(PIN_SD_CS, OUTPUT);
    digitalWrite(PIN_SD_CS, HIGH);
    Tft.TFTinit();
    Sd2Card card;
    card.init(SPI_FULL_SPEED, PIN_SD_CS);
    Tft.drawString("Initializing", 0, 0, 3, WHITE);
  if (!SD.begin(PIN_SD_CS)) {
    Tft.drawString("Failed", 0, 30, 3, WHITE);
    return;
  }
  Tft.drawString("Success", 0, 30, 3, WHITE);
```

```
    readFile();

}

void loop() {
    // a point object holds x y and z coordinates.
    Point p = ts.getPoint();

    //map the ADC value read to into pixel co-ordinates

    p.x = map(p.x, TS_MINX, TS_MAXX, 0, 240);
    p.y = map(p.y, TS_MINY, TS_MAXY, 0, 320);

    // we have some minimum pressure we consider 'valid'
    // pressure of 0 means no pressing!
    if (p.z > MinimumPressure && p.z<10000) {
        if(p.y < totalTextHeight && p.x > rightMargin){UpDown = -1;}
        if (p.y > screenHeight - totalTextHeight && p.x > rightMargin) {UpDown=1;}
        delay (500); // Prevent bounce. may vary time
        }
      if (UpDown!=0) {
       if (p.z == 0) {
         startItem=startItem-UpDown;
         if (startItem<0) {
           startItem = 0;
           } //Do not let start item be less than first item
         if (startItem>numItems-maxItems) {
           startItem = numItems -maxItems;
           }
         UpDown=0;
         displayText();
         }
      }
}

void readFile(){
  File dataFile;
  dataFile = SD.open(fileName);
```

```
  if (dataFile) {
    numItems = 0;
    // read from the file until there's nothing else in it:
    while (dataFile.available()) {
      data[numItems] = dataFile.readStringUntil('\n');
      numItems = numItems + 1;
    }
    dataFile.close();
    displayText();
  }
  else {
    // if the file didn't open, print an error:
    Tft.drawString("Cannot read",10, 90, 3, RED);
  }
}

void displayText() {
  Tft.fillRectangle(0, 0, screenWidth, screenHeight, BLACK);
  Tft.fillRectangle(rightMargin, 0, screenWidth, totalTextHeight, BLUE);
  Tft.fillRectangle(rightMargin, screenHeight - totalTextHeight, screenWidth, screenHeight, BLUE);
  for (i=0;i<maxItems;i++) {
    if (i + startItem < numItems){
    int len=data[i + startItem].length();
    char DisplayableString[len];
    data[i + startItem].toCharArray(DisplayableString, len);
     Tft.drawString(DisplayableString, 0, i * totalTextHeight, textSize, textColor);
    }
  }
}
```

Sketch 12.3

This basically combines the text reading sketch with the scrolling text display sketch. Naturally, there are some changes necessary to each of these sketches to merge them and create the new functionality.

The first change is in the readFile subroutine. Where

previously it had sent the strings it loaded from the file to the serial port, in this sketch it stores them in a string array. That string array was declared in the beginning of the sketch with
String data[50];
The selection of 50 as the top of the array was arbitrary. You should dimension the array to suit the number of lines you expect in the text file. In the readFile subroutine, the line
data[numItems] = dataFile.readStringUntil('\n');
stores the string read from the file in slot numItems of the data array. The value of numItems is set to 0 before any strings are read in. You may note that it was also set to 0 when it was initially declared at the beginning of the sketch. The reason for resetting it to 0 within the readFile subroutine is to allow you to call the readFile subroutine repeatedly with different file names from within the program to load in different text if needed. Of course, if you actually wanted the new text to be added to the list of text already in the array, you can delete the numItems = 0 from the readFile subroutine. After one item has been loaded into the data array, the value of numItems is increased by 1, so the next string read in from the file will be a separate item in the data array. After all the strings in the file are loaded, the file is closed by dataFile.close() and then the displayText subroutine is called.

    Note that the code for reading the text from the files uses string type arrays and the Tft.drawString command used to display characters on the screens uses char type arrays. Therefore, in the displayText subroutine, you will need to convert the string variables to char arrays in order to use the drawString command. This is similar to the situation we had with the numeric keypad in Sketch 7.1, where we needed to use strings to manipulate the string of numbers being displayed but needed to convert that string to a char array to display it. In this case, when we want to display the string array item data[i + startItem], we first get its length with
int len=data[i + startItem].length();
Then we create a temporary char array with
char DisplayableString[len];
Then we store data[i + startItem] in DisplayableString with
data[i + startItem].toCharArray(DisplayableString, len);
Then we use the Tft.drawString command to display this string of text.

    In the previous sketches, the name of the file was hard coded into the sketch. It might be nice in some circumstances to be able to select the file to view or otherwise use from a menu. The following code demonstrates this.

```
#include <SPI.h>
#include <SD.h>
#include <TFTv2.h>
#include <SeeedTouchScreen.h>

// SD card variables and constants
const int PIN_SD_CS = 4;
File root;

// Menu variables
int screenHeight = 320;
int screenWidth = 240;
int MinimumPressure = 50;
int textSize = 3;
int textSpacing = 5;
int textColor = CYAN;
int selectedTextColor = WHITE;
int printColor;
int totalTextHeight = 7.5 * textSize + textSpacing;
int maxItems = screenHeight/(totalTextHeight);
String menuItem[20];
int numItems=0;
int i;
int lineTouched;
int Selection=-1;
boolean SelectionMade = false;
TouchScreen ts = TouchScreen(XP, YP, XM, YM);


void setup(){
  // Open serial communications and wait for port to open:
  Serial.begin(9600);
   while (!Serial) {
    ; // wait for serial port to connect. Needed for Leonardo only
   }
  Serial.print("Initializing SD card...");
```

```
  pinMode(SS, OUTPUT);
   digitalWrite(PIN_SD_CS,HIGH);
   Tft.TFTinit();
   Sd2Card card;
   card.init(SPI_FULL_SPEED, PIN_SD_CS);
  if (!SD.begin(PIN_SD_CS)) {
   Serial.println("initialization failed!");
   return;
  }
  root = SD.open("/");
  loadDirectory(root);
  Serial.println("initialization done.");
}

void loop()
{
   Point p = ts.getPoint();
   p.x = map(p.x, TS_MINX, TS_MAXX, 0, 240);
   p.y = map(p.y, TS_MINY, TS_MAXY, 0, 320);
   if (p.z > MinimumPressure && p.z<10000) { // If pressure on screen
     lineTouched = p.y / totalTextHeight;
       if (lineTouched <= numItems-1) {
         Selection = lineTouched;
         displayMenu();
         SelectionMade = true;
       }
    }
     delay (250);
    if (SelectionMade && p.z==0) {
       // Take some action on selection
       Serial.println (menuItem[Selection]);
      SelectionMade = false;
      Selection=-1;
      displayMenu();
    }
}

void loadDirectory(File dir) {
  dir.rewindDirectory();
```

```
  while(true) {
    File entry = dir.openNextFile();
    if (! entry) {
      break;
    }
    menuItem[numItems] = entry.name();
    numItems++;
    entry.close();
  }
  displayMenu();
}

void displayMenu() {
  for (i=0;i<maxItems;i++) {
    if (i< numItems){
      if (i == Selection){
        printColor = selectedTextColor;
      }
      else {
        printColor = textColor;
      }
      int len = menuItem[i].length()+1;
      char DisplayableString[len];
      menuItem[i].toCharArray(DisplayableString, len);
      Tft.drawString(DisplayableString,0,i * totalTextHeight, textSize, printColor);
    }
  }
}
```

Sketch 12.4

The subroutine loadDirectory loads the directory of files into a string array menuItems. This process can be much more sophisticated, allowing you to search through folder trees for files, but for the sake of simplicity I will assume that you have all your files in the root directory.

The variable root is declared at the beginning of the sketch as type File. The variable dir is declared within the subroutine declaration itself. The variable numItems (the number of items in the menu) is set to 0.

The command dir.rewindDirectory() sets the directory to the first file. We then have a loop that is guaranteed to continue until stopped because the test condition is set to true. Within this loop, the line

File entry = dir.openNextFile();

creates a variable named entry as type File and loads it with the next file in the directory. If there is no next file (all files have been found), the if statement breaks out of the while loop. If there is a next file, the statement

menuItem[numItems] = entry.name();

loads its name into the menuItems array at position numItems. The variable numItems is then increased by 1. Thus, the first item in the menu is menuItem[0], the second is menuItem[1], and so on. After all the file names have been added to the array and the while loop ends, the command displayMenu() displays the menu on screen.

You may only want to display certain types of file names, such as bitmap files. If so, the following change to the loadDirectory subroutine will allow that.

```
void loadDirectory(File dir) {
  numItems = 0;
  String temp;
  dir.rewindDirectory();
  while(true) {
    File entry = dir.openNextFile();
    if (! entry) {
      break;
    }
    temp = entry.name();
    if (temp.endsWith(".TXT")) {
      menuItem[numItems] = temp;
      numItems++;
    }
    entry.close();
  }
  displayMenu();
}
```

This is basically the same as the previous loadDirectory, with a few additions. It defines a string variable temp, and assigns the entry name to it after the entry variable has been given the file name by dir.openNextFile(). Then the line if (temp.endsWith(".TXT")) checks to see if the file name ends in ".TXT". If it does, this file name is added to the menu array. You can, of course, substitute any file extension for ".TXT".

Appendix

Downloading the Files and Contact the Author

You can download the files in this book from the following links, as of the writing of this book:

https://github.com/DavidLeithauser/Touch-Screen-Book-Files

http://LeithauserResearch.com/TouchScreenFiles.zip

These are ZIP files that you can unzip into a complete folder structure. The main folder is TouchScreenFiles. Within this main folder are folders with names like S3P1, which stands for Sketch 3.1. Each folder contains a file with the same name and the extension .ino, ready to be loaded and used. There is one folder for each named sketch. I have not provided copies of the partial sketches, like subroutines that I list separately, but such pieces of code are almost always included in a full sketch within the book, and therefore can be found in the files you can download.

Once you unzip this archive, you can move the entire TouchScreenFiles folder to the
C:\Program Files\Arduino\examples
folder to make it easy to use. Once you do that and run the Arduino IDE program, you should see the folder in the examples list.

If you have any questions about this book, I can be reached at Leithauser@aol.com.

www.ingramcontent.com/pod-product-compliance
Lightning Source LLC
Chambersburg PA
CBHW070256190526
45169CB00001B/431